Backyard

BOATBUILDER

Backyard BOATBUILDER

How to Build Your Wooden Boat

John Welsford

REED

Published by Reed Books, a division of Reed Publishing (NZ) Ltd, 39 Rawene Rd, Birkenhead, Auckland. Associated companies, branches and representatives throughout the world.

ISBN 0 7900 0312 0
First published 1999

Cover and text designed by Michele Stutton

Foreword

This book is based on a compilation of material that shares a common theme. Some of the material was developed from John's design commissions and research, while other pages have been drawn from the many practical magazine articles he has produced in the past.

As a magazine editor drawing on John's designing, boatbuilding and writing skills, I generally processed his features last, not only because he pushed his deadlines (!) but because his concepts were achievable, founded on common sense, and sometimes just a little provocative. He always wrote good copy that I could trust and enjoy.

Often I found myself identifying with his designs and the thought processes that shaped them. It was not unusual to be transported from my 'fluoro'-soaked desk to another world, reliving past times that had been good for the soul, or dreaming of new maritime adventures that could be accomplished within the tight constraints of my suburban existence. Even when describing a rowing boat, John could impart an aspect of recreation and adventure that few would associate with such a craft.

It was impossible to categorise John's work. Initially I thought of him as a traditionalist, for many of his sailing and rowing boats have that look and function. Sometimes there was the hint of the alternative, and occasionally, when specifically asked, he gleefully produced a design that embraced the radical. In the end I accepted that he was comfortable working with a range of concepts and mediums, many of which can be identified within this book.

John's preferred materials for construction — plywood, timber and epoxy composites — with the emphasis on plywood and timber — are often reflected in his designs. After working in the plywood and timber industries and guiding numerous amateur builders through their first projects, he is adept at utilising these materials to best advantage.

Regardless of the design or construction task at hand, John applies a number of guiding principles. The construction method must be easy, economical and efficient. The boat must be sound, seaworthy and ergonomic. And the final result must be fun and enjoyable.

So, if you appreciate sailing or power, rowing or paddling, have a bent towards home construction, appreciate free thinking, and enjoy a good read, this may very well be the book for you.

I know it will sit well on my bookshelf.

Geoff Green

Acknowledgements

Looking back on how a book such as this came into being, and finding someone else to share the blame, isn't easy. The designing and building are a part of it. Those people who built boats were to me, a terminally afflicted boating nut, a special breed of superhumans, only bettered by those who actually designed them. While as one who reads, on average, a book of some sort every two days, authors to me are people who sit on the other side of God from boat designers. To be introduced to any one of the three would strike me speechless. Those who know me will no doubt find that hard to believe.

During my childhood Dad never laughed at my attempts to build my own toys. He encouraged my dreams, and by his own example showed me that, while a body has physical limitations, one's mind can be as free as its owner wishes.

Granddad Leslie took me out fishing in his little clinker dinghy every summer holidays. He left me under the watchful eye of boatbuilder Philip Lang, ensuring I had the bug so badly girls and motorbikes would only be allowed to interrupt the affair for a short time.

Along the way there was Colin Peffers, an extraordinary teacher, who made up for a lot of shortcomings in a school system that didn't fit me too well. Another schoolteacher, now retired, is Neil McLeod, who showed me that it is never too late to live, that failure is in itself an opportunity to learn, and that failure should never be feared.

There was an 'old sea dog' in the form of Jack Salter, who encouraged my first attempts at design. He taught me that I was at least as good a person as anyone on this earth, and that 'impossible' really meant 'let's have another look at this tomorrow, we might have thought of something by then!' There is yet another teacher, this time friend Bob Jenner who has built three of my designs — a man whose example in life I humbly follow.

A vote of thanks goes to my friends at the Traditional Small Craft Society, many of whom paid me the ultimate compliment of building one (or, in Frank Bailey's case, two) of my boats; to Peter McCurdy, who took some of my early writings and made them fit for human consumption; and last but not least, to Shane Kelly and Geoff Green, who, while editors of *Sea Spray* and *Boating World* magazines, helped me improve my writing, and who always had confidence in my ability to deliver.

To all my friends everywhere: you are all a part of me, and each of you has contributed to what you read in this book. Thank you all.

John Welsford

Contents

Methods and Equipment

Chapter One

From Simple Beginnings . . .

An expert is just a beginner with experience.

There is much to be had from small craft: a Tender Behind, a Joansa, my wife Jan's kayak, and a truly traditional pulling boat by Captain Jim Cottier, at the very head of the Puhoi River.

Messing about in boats is, for some, a lifetime affliction, and so it has proved to be for me. At the tender age of eight or nine (age didn't seem to matter much then, at least until girls arrived on the scene a few years later), I was building canoes from sheets of corrugated iron, and giving my Mum grey hairs as I explored the estuarine shores of the upper Waitemata Harbour where the family farmed.

These 'boats', although much the same as those built by generations of small boys, taught me a lot. They improved my swimming for one! I learnt that a narrow boat with a dead straight bottom would not turn but would toss me in the tide if I so much as looked sideways. Pushing the

sides out wider gave more stability and more rocker (fore and aft curve in the bottom), but the boat could not be paddled straight without some form of skeg or keel to slow down the turn; all good stuff, and a grounding which has served me well — I don't fall in so often these days!

Boats have remained a part of my life, along with excursions into model aircraft which taught me a lot about plans and lightweight structures, and motorcycle road-racing which taught me that there had to be a better way of satisfying my need for thrills than sliding down the road on my behind at a hundred miles an hour! There were sailing dinghies, including an 18 foot mullet boat which I note is now referred to by its owner as an 'historic' vessel. I thought it was more 'hysterical' and was only on the scene because I couldn't afford anything better. Later on I spent a lot of time on racing keelers.

In spite of my involvement in other interests there was much sailing on other people's boats. When my employer (not a fan of motorcycle racing) put it to me that my imminent promotion was dependent on my not being involved in a 'hazardous' sport (I was getting tired of falling off anyway), a 21 foot trailerable yacht that I had built myself rekindled a flame that has had me building and designing boats ever since.

I raced my yachts a lot and was moderately successful, but after years of thrashing around the buoys I was fortunate enough to discover small boat cruising, and I've been exploring the coasts and rivers in a variety of small sailing and rowing boats ever since.

As my experience grew I found myself wanting to share the fun of what turned out to be a cheap and interesting way of cruising — boats that didn't have to cost a fortune and that were a great deal more seaworthy than most would imagine; boats an ordinary guy (or guyess) could build without special skills or tools; and boats that would look so salty that they would make the onlookers' eyes rust in envy!

I've been in the designing business for fourteen years now. From a small start producing designs mostly for myself and the occasional trusting friend, my plans trade has grown to a substantial business, with examples of my boats scattered throughout New Zealand and Australia, as well as pockets of them in several other countries.

Reaction to these boats, with their semi-traditional styling, practical rigs and strong but lightweight construction, has been very positive, suggesting that many people would get stuck in and build themselves one if only they had the confidence.

My old friend Jack Salter told me, 'An expert is just a beginner with experience.' You'd better believe it — if you make a start, and keep steadily at it, you'll learn a lot, suffer a few frustrations and have a lot of fun, but in the end you'll have a boat of your own. That's how I did it, and even now, after building fourteen boats in twelve years (I have to work for a living which slows me

down a bit!), I still get the same thrill as always when I build, or launch a new one.

The first time out on the water in a boat built with one's own hands is wonderfully satisfying, experience. Having started with a set of plans, a pile of materials and an empty garage, it is just great to bring the boat out into the light of day for the first time. Friends who came and sat, offering 'helpful' comments while I fought with a springy plank, now look on in envy as this shiny, graceful butterfly emerges from her chrysalis.

The suburban backyard builder's 'boatyard' — a Light Dory *and* Daniel's Boat *going together in my garage.*

The first feeling of life as she lifts to those little waves near the shore, as she heels to the first puff of wind or surges away under oars, is a real reward for those many hours that went into creating her, your boat!

Building it yourself can be rewarding. It can also be frustrating and demanding, to such an extent that even after investing much money and time projects are abandoned. Taking on too big a boat or one not suited to the builder's needs or resources can lead to real problems. Doing it yourself should not be like this.

In this book I have endeavoured to illustrate the means and methods that have enabled me to turn out one or two small craft each year, enjoying the process as well as using my time in the workshop as a break from the rat race that never seems to get us ahead of the bills. The boats have proven to be good investments, typically fetching about double the materials' cost when sold to make room for the next project.

Paul and Cath Cooper's Demelza. *At 8.3 m (27 ft 4 in) length and 2.9 m (9 ft 6 in) beam, about as big as you'd want to cope with on a spare-time basis. She is the product of about five years of hard work.*

Boatbuilding without undue stress requires one to treat the project as a hobby in itself, one with its own satisfactions and rewards.

It helps to allocate time to the project on a planned and regular basis, and pays to do the same with the family so they don't feel abandoned in favour of the 'thing' in the shed. Try to keep them involved where possible and share the dream with them. It is often family pressure that finally kills the bigger projects, pressure created when a spouse does not share the interest that has led a partner to build a 'bigger boat'.

Although this book has some quite big boats featured, which I always think of as my 'dream ships' (these dreams are, however, intended to be relatively achievable ones), the smaller boats that are the main focus of this epistle are within the means of most people in terms of skills, money and time required. For example, the cruising dinghy *Rogue* can be built, rigged and sailing for about three weeks' ordinary wages, and most of the eighty or so built to date have taken their first-time builders between four and six months to complete.

My own *Rogue* took me on extensive coastal cruises and was raced with surprising success, as well as being a very nice daysailer and being useful for a bit of fishing on the side. She did not break the bank, was easily handled by one while still being able to carry a group of friends, and all in all was a prime example of the old saying, 'The amount of fun you get from a boat is in an inverse ratio to the amount of money you have invested in her.' How often do you see beautiful big ocean-going yachts lying forlorn and unused on their moorings?

Of the designs featured in the second section of this book, there are a range of types, mostly modest in size, that will fill the real needs of many boaties. Note that I say 'real' needs. Many of those who start off building a bigger boat are not only biting off more than they can chew but will, when (and if) completed, end up with a boat much bigger and harder to handle and maintain than they really need — a recipe for dissatisfaction and stress.

Do be realistic about your needs. Choose something that will be well within your resources and will fulfil only the essential parts of your planned use. Dream, but have practical dreams. You'll find that these little boats are much more capable than the modern myth would have you believe. In fact, I have taken my small cruisers much further from home than I ever took my keeler, and that was before I started to trail them to ports even further away.

We visited interesting places closed to deeper draft boats, found that people were much more friendly to travellers in little boats, and found that we could be very comfortable (and secure) in our tent ashore each night.

When building your boat start with a minimum of tools. After you've been at it for a while you'll have a better feel for the gear needed. Start with a few good quality items; read the chapter on tools before splashing out. Don't spend too much, and don't get depressed about the list of items you appear to be without.

Buy top materials for your boat; anything less devalues your valuable labour. It can also be much harder to achieve a good-looking boat with cheap plywood; the same goes for paint and the rest of the materials. Remember that this is fun: looking at a boat that has given you much satisfaction in the building and that attracts the admiration of the public when out sailing on the briny is very good for the soul!

My friend Wayne Chittenden and I having fun in Hobo, an early small craft design. Photo: Peter McCurdy

Chapter Two

Understanding the Material

What Is Plywood?

In order to utilise a material we should first explore its characteristics. Plywood is a panel made from a number of layers of wood veneer, pressed and glued together to produce a sheet of laminated wood with similar strengths in two of its three directions.

Note the word 'similar'. A normal wooden board has a high compressive and tensile strength along the grain and poorer strength across the grain (Figure 1). The difference gets less pronounced as the number of 'plies' or layers of veneer opposed to each other — 'long veneers' (longbands) and 'crossbands' — increases. In three-ply there are two outer long veneers and one interior crossband. In five-ply, the ratio is three longbands and two crossbands, and so on. Generally, seven-ply has almost the same strength in both directions, five-ply is close enough for most purposes, while three-ply requires some thought to be given to the orientation of the outer ply's grain direction.

It is rare to use more than seven-ply, and two-ply is not normally employed for marine or structural use except when building up a fancy curved component, but even then there will be two or three layers, effectively making four- or six-ply. This is probably the only time you will come across even numbers of veneers. Also rare is plywood with all or mostly longbands and few or no crossbands. There are many variations on plywood or laminated veneers, but most fall outside the scope of this book, so you can assume that any type we mention from here on can be used for building your heart's desire.

Another important strength of plywood is

Plywood starts as an uninspiring sheet. Here it is on its way to becoming a shapely boat.

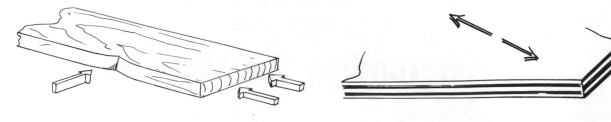

Figure 1: A solid wooden plank is much stronger in both compression and tension along the grain than across the grain.

Figure 2: Plywood makes a very rigid bracing medium.

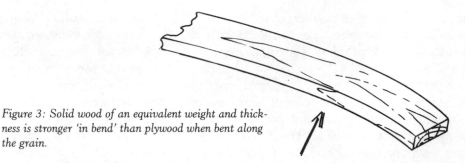

Figure 3: Solid wood of an equivalent weight and thickness is stronger 'in bend' than plywood when bent along the grain.

diagonal, or 'bracing' strength (Figure 2). The stiffness of a plywood panel across the 'angle' is an important advantage, and a clued-up designer will engineer a vessel to take advantage of this, producing a boat incredibly strong for its weight. Also worth noting is that converting a traditional, solid-timber plan to plywood should involve a complete redesign, retaining only the shape, and even that could need changing as the boat may be a lot lighter.

Solid wood, with its unidirectional grain orientation, has the advantage when bending that all of its bulk is being stressed in its strongest direction (Figure 3), while plywood has the disadvantage that two out of five veneers (40 percent of its volume) are stressed in their weakest direction.

One great virtue of plywood is its stability. When adhesive is used to lock together layers of wood veneer with opposing grain directions, it produces a resistance to the dimensional change which is caused by an increase or decrease in moisture content.

Wood swells mostly across the grain as moisture content is increased, but by locking pieces with opposing grain together the change in dimension is greatly reduced. Also assisting this stability is the almost completely waterproof glue line between each layer, so even the thickness should not increase greatly with wetting of the panel.

Note that there will be some swelling or shrinkage in some circumstances. I recall one proud owner who had made a superb job of a sailing dinghy. He had carefully filled every defect and blemish before painting, had buffed the paint to a mirror finish, and taken endless pains to ensure perfection. Much to his chagrin, shortly after he launched her the filler seemed to shrink, showing every nail and screw position.

What had happened was the boat had been moved from a dry and warm workshop to an outside carport. Over a few months the ply had increased ever so slightly in moisture content and swelled just enough to show the filler, which was more stable. Note that paint is not a perfect barrier against water!

Plywood has some problems in the area of decay resistance. Unless edges are adequately sealed, crossbands will soak up water like any end grain, and this can produce delamination and rot which may travel extensively through the panel before becoming evident to the hapless boat owner. There have been a number of failures, some quite spectacular, in older small planing powerboats that have probably been

Well on its way — Houdini *partly planked up. The self-supporting 'monocoque' possibilities of plywood are shown here — note the lack of framing.*

caused by deterioration of this nature. Ingress of water into any end grain should be carefully guarded against by using proper building procedures. The most effective method of doing this is to carefully coat the edges of a component with epoxy resin before fitting. Two coats are better than one!

Plywood which has interior (core) veneers of different wood to the outside (face) veneers seems to be particularly prone to problems of this type, and in my experience should be avoided.

However, due to the careful selection of logs and the grading of veneers, most high-grade (certified marine BS 1088) plywoods are free of sapwood or soft spots that present opportunities for rot spores to begin their destructive work.

On the flat a plywood sheet is not particularly strong. However, when 'pre-stressed', that is held in a bent or curved position, it is very much stiffer, stiffness being resistance to movement, as opposed to strength which is resistance to breakage.

Hence we have a material which:

1. Comes in large sheets of fairly standard sizes — 2400 x 1200 mm or 2440 x 1220 mm.

You can't just find a longer log and cut a plank but with care and epoxy resin glues, sheets that are too short for the job can be spliced or scarf joined into longer lengths.

2. Is comparatively uniform in strength on two axes.

3. Is easily bent in one direction.

4. Is stiff in two other directions.

5. Is dimensionally stable.

6. Can be very durable if selected and used with care.

7. Can be easily worked by anyone with basic woodworking tools and skills.

8. Nails, glues and paints even better than the 'real' thing.

9. Is a lot less harmful to the builder's health than some of the hellish brews used to make our high-tech plastic boats.

10. Can be readily obtained at a remarkably consistent quality and, in comparison with most alternatives, is quite reasonably priced.

Marine ply, BS 1088, is solid, has no voids (gaps in the core) or defects, is marine bonded and guaranteed. It is the best to use for boatbuilding, and is also top of the line for price.

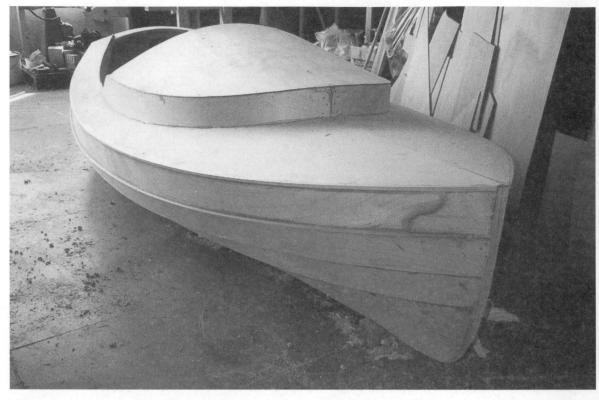

Sheets of plywood can be formed into quite sharp curves to form parts of a boat.

Generally, if the ply has a dark red glue line it is glued with phenol formaldehyde (a waterproof glue); provided the veneer is of a naturally durable timber species, or treated, and the sheet looks clear of defects you could use it — with caution or advice. If in doubt, insist on BS 1088 stamped material (not available in construction plywoods), particularly if building more expensive or surveyed vessels.

Construction ply, made from pinus radiata (*Pinus insignus*), Douglas fir or sometimes ponderosa pine, is becoming more and more popular with improvement in grades, treatment methods and competitive pricing.

Construction ply is easily available, is sometimes already treated for decay resistance, and can be obtained in a wide variety of grades. Largely because of its price, it has built up quite a following among amateur builders, while professionals use it for bulkheads and the out-of-sight areas in their building.

The major virtue of construction ply is price, and used correctly, its drawbacks are easily overcome.

Treatment

If using treated ply, always select pressure CCA (copper, chrome and arsenate, otherwise known as 'tanalised™') treated ply, characterised by its dull green colour and no smell other than that of timber or pine resin. Do not put the offcuts on your fire, or the sawdust on your garden, and — most importantly — don't get the sanding dust in your lungs!

Construction ply is usually CCA treated to 'H3' (moderate hazard) but in thicknesses to 12.5 mm it can be obtained in 'H4'. This is not marine standard but it is enough to prevent rot spores getting a hold and will give marine worms severe indigestion! Note that it is very difficult to treat plywood to the higher standards, and as far as I know 'H5' (marine treatment) is not available in construction ply.

Take care with cut edges, the treatment does not extend far into the core because the glue lines stop the transmission of the preservative and it may not go in more than 100 mm or so from the edge of the sheet. There is another treatment, an oil-based one called LOSP (Light

Oil Solvent Preservative) which will not glue. It is a very light but bright green, as opposed to the dull finish of tanalith™, and noticeably smells when a packet of ply is opened. Do not use it!

Watch for core voids (gaps in the crossbands or core longbands), overlapping crossbands, and knot gaps in the edges of core veneers. Look for one-piece crossbands or at least close-fitting joins. Deep tearouts in the face grain should be rejected, as these are almost impossible to sand out and are hard to fill.

Sheets which appear curled, cupped or propeller twisted should be avoided, particularly in five- or seven-plies, as they are extremely hard to straighten out. Look for plywood brands which use one-piece crossbands as these are better suited to marine use.

Grades of Construction Plies

'Case Grade' — rubbish. This is what you build your temporary shed out of.

'CD' — one 'C' face, one 'D' face. Cheap and usually looks it. Usually unsanded, this might suit a quick-and-dirty skiff.

'CPD' — one 'C' veneer, knots and splits filled with epoxy filler sanded at least on this face, and one 'D' face. This is getting into more

usable material — sometimes it is worth picking over the stack if the yard manager will allow you. Do put it back tidily afterwards.

'BD' — the 'B' face should be nearly perfect, and sanded. This may be a good choice, but remember to put the 'D' face outwards where you can fill and sand the defects; knotholes and checks (small splits) on the inside are rot traps and are hard to get at to fill, whereas outside you'll be filling and sanding off anyway.

'BB' — two near-perfect faces. This ply is the way to go but is sometimes hard to get. Many merchants will not stock it but it may be worth phoning around to find some. Sanded, and usually has high-grade cores.

'AN' or 'A' — top-of-the-line construction ply. Expensive, but good stuff if you can get it.

Finishing construction ply is a problem, as the face veneers are prone to checking when exposed to sun and weather. I have had fair results finishing plywood boats with epoxy-glass cloth skinning, or with epoxy saturation and/or white enamel house paint over an oil-based base coat and high-build undercoat primer. However, the surface is not as stable as some other plywoods, and for outside use it should never be finished in a dark colour unless

Even a quite simple plywood boat can be graceful and a thing of pride. In this case, the boat is Colin Quilter's beautifully built Seagull. Photo: Colin Quilter

Tender Behind's hull is essentially nine pieces of plywood, each one quite easily bent, almost floppy. By holding each piece in a curve the assembly is incredibly stiff.

fibreglassed. Clear finishing of any kind is out for exterior use.

Interiors done in construction ply can be finished with any normal interior marine finish — varnish, enamel or polyurethane, with the proviso that care is taken to apply sufficient 'sealer' to prevent the porous surface from endlessly soaking up your finish coats and leaving rough patches in your smooth glossy paint.

A note on plywood qualities in general: your labour is valuable, and your boat is an expensive and labour-intensive project, in which — in the more ambitious vessels at least — the cost of the plywood is only a small proportion of the total. Buying top-quality materials does not add hugely to the total cost, while an inappropriate grade of cheap plywood may so devalue your finished boat that you'll regret your choice for ever. Do buy quality that suits the intended standard of your boat — it may return you a much better result for a comparatively small difference in cost.

Chapter Three

Adhesives

As the bank robber said to the teller, 'This is a stick-up!'

In 'days of yore, hearts of oak, wooden ships and iron men, oakum tar, naval brass' and all that, glues did not feature at all in keeping together the myriad pieces of wood that made up even a small craft.

During the intense technological race of World War II, complex, highly stressed wooden structures, such as the famed DeHavilland Mosquito bomber, were developed using 'new technology' in the form of urea formaldehyde and later phenol formaldehyde glues.

The impact of these wood glues was enormous. Plywood became something more than just a decorative panel, and wood was no longer limited to what could be cut from a log, and nailed or riveted to a framework, which in turn depended on metal fastenings. Most importantly, they made possible a 'style' of boatbuilding more readily achievable by beginners.

The skills necessary to build a smooth, finished (carvel planked) hull from solid planks on steamed frames to a good standard are no longer common, and are not easily learned in the course of building one small boat for your own use.

However, the techniques developed to take advantage of modern adhesives, while still requiring commitment, are not beyond the skills of the ordinarily handy person. Sheet plywood, lamination of large structural members, and scarfed joints as strong as the unbroken timber, have effectively turned big projects into a series of small ones, each one not difficult in itself. Some of the big baulks of wood used in traditional methods are difficult to work, difficult to obtain, and difficult for a sole worker to handle in a confined workplace that often has a minimum of equipment.

Please, traditionalists, do not see me as dismissing the tried and true methods of time gone by. Rather, see me as advocating methods that will enable people, who may never be dedicated students of tradition, to get their heart's desire on the water with the minimum of time and expense.

Most glues used in boatbuilding are so strong that in a properly made joint, the wood will break before the glue fails. In this chapter I will run through a number of glue types with a brief explanation of what they are and how I find them in use from a practical point of view. Note I include both PVA and hot melt glue to point out that, while they are common, they are not suitable for boatbuilding.

PVA (Polyvinyl Acetate, or White Joiner's Glue)

This glue is not waterproof or even moisture resistant. It's not gap filling and, in many applications, is not particularly strong. I don't use it anywhere. Don't let the bod on the counter at your local DIY store convince you otherwise.

However, there are some waterproof PVA glues beginning to appear on the scene. The ones I have seen are bright yellow (don't bet on one just because it is that colour — do your homework), and seem to be a very good and relatively non-toxic adhesive (don't bet on that either — work clean!), well suited to interior joinery. If you want to try this out, write to the manufacturer who will supply a technical specification that will outline the test results for the type under a wide range of conditions, including humidity and moisture. If in doubt, check it out!

Cyanoacrylates (Superglue)

This is expensive and has very limited use; it doesn't work well on porous surfaces. However, I do keep a tube on hand for use on plastics, light metalwork and on some electrical gear (not where any heat is likely). It is wonderful stuff — where it is appropriate — but do be careful to keep it off your skin, and be particularly careful to keep it away from your eyes.

Urea Formaldehydes
(Aerolite 300 or Similar)

This is where modern adhesives began. The days of glued wooden aircraft truly started with the development of UF resins, hence the name Aerolite, and led to the development of modern glue systems.

UF glue is water resistant but should not be used where immersion in water is intended. It is great for interior joinery where the completely transparent glue line is an advantage on the varnished trim. The ability to keep resin (powder and water) mixed and apply the catalyst only at the time of assembly is a help, but do remember it requires a very close-fitting joint and positive clamping pressure. I like this glue, and use it for interior trim in larger boats.

Melamine Urea Formaldehyde
(Aerolite 308 or Similar)

This is extensively used in industry. For example, much of the 'finger-jointed' timber in modern homes is glued with this. It is water resistant, much more so than the unmodified urea formaldehydes. I would go so far as to suggest that it is waterproof but not boilproof, so if you intend sailing your dream craft in boiling water, don't use it. This is not as silly as it sounds; as I write I am about to move to Rotorua, an area of many lakes and much geothermal activity. Some of the lakes have patches of pretty hot water, so you can bet that I'll be staying clear of the few really hot spots!

Aerolite 308 is advertised as being 'gap filling'; that is, it includes a thickening agent and is non-shrinking. The gap-filling ability does not cope with 6 mm gaping chasms, but perhaps 1 mm or so is okay.

The advantages of UF glues are also present in MUF, and both can be batch mixed with catalyst and used in bulk, as in two-skin ply or cold moulding. The well-known designer and builder, John Spencer, was a great advocate of UF glue, and many of his craft, some now with many thousands of hard miles under their keels, are still around and are still structurally sound.

Phenol Formaldehyde (Aerodux or Resorcinol Glue)

This is the stuff that holds plywood together —

waterproof, boilproof, very strong, easily handled and, in common with UF and MUF glues, seems not to be particularly temperature sensitive. However, don't take this statement as a blanket licence to work outside the manufacturer's recommendations.

Resorcinol or Aerodux are characterised by dark red resin which leaves a near-black glue line, and will sometimes stain the area around the joint, causing an unsightly discolouration if using varnish.

This glue generally comes as a liquid resin and powder hardener (be careful with mixing ratios), and has quite a good 'pot life' (time from mixing to being usable). It requires a close-fitting joint as it is not gap filling, with firm clamping pressure required.

I find the PF resins very good for work where my usual tendency to be in a tearing rush with my building has not left big gaps for the glue to fill. Clean-up is with water, as long as it is within the pot life time.

This is a very good wood glue and — until the common availability of epoxies — was the industry standard. It is still in widespread use, and for some styles of boatbuilding is, to my mind, one of the best.

Note: those people concerned about the possibility of allergic reactions to epoxy systems may find that it is possible to modify the construction methods of some boats intended for epoxy construction to the use of 'red glues'.

Epoxies

These are the modern miracle and the amateur builder's best friend, but like most things that are exceptionally good, there are some catches.

Safety precautions must be observed. Some people are very sensitive to the chemicals in epoxy resins and their hardeners; not having any ill-effects from using the stuff in the past does not mean that you won't have a problem with them in the future. The problem seems to be cumulative and once evident may reappear with even the minimum of exposure.

When used as directed, epoxy resin, mixed with the various 'extenders' that convert the liquid resin into glues and fillers of high or low density as needed, is a wonderful tool. Very strong, gap filling like nothing else, and tremen-

Some of the chemicals in adhesives don't do you any good! Gloves are cheap, and you'll quickly get used to wearing them.

Epoxy resin in use — note the pump dispensing system. To the mixed resin and hardener is added the appropriate powder to make glue, filler, or fairing compounds. Note the gloves!

dously versatile, an epoxy resin dispensing system is not just glue but is the key to a very user-friendly boatbuilding system that cannot be matched by any other method.

These resins cure at a wide range of temperatures; do read the instructions though, as too hot cures the thermosetting plastic too quickly while too cold can prevent a cure.

I believe that many users treat the safety aspect too lightly; we do not yet know the full implications of exposure to some of the chemicals involved.

I use epoxy resins for almost all of my boatbuilding, as they suit my sloppy and impatient carpentry. They also provide me with various glues, and high- and low-density fillers that enable me to radius corners and joins in such a way that the boat looks almost like a unified organic whole, rather than an accumulation of pieces.

Hobo on the Manukau Harbour. This water-ballasted 4.9 m (16 ft) craft, with her cat yawl rig, was one of my all-time favourite boats. She has only a few fastenings and is an almost entirely glued structure.

The Gougeon Brothers of WEST ™ fame have extensively researched and popularised a boat-building 'system' based on epoxy resins. They promote the 'saturation' of each individual piece of wood as it is included in the structure, stabilising the material as well as fastening it into place. Their book *The Gougeon Brothers On Boatbuilding* is a real treasure, and is valuable reading for anyone who wants to explore modern wooden boatbuilding.

Don't forget Murphy's Law of Boatbuilding: 'Any mistake you make will not be discovered until the glue has set.' As an antidote I give you Welsford's Law of Boatbuilding: 'The mistake that cannot be rectified by liberal use of epoxy and fibreglass hasn't been made yet!'

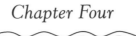

Chapter Four

Tools

It's easy to spend too much …

A hammer for every job! However, the light claw hammer is really the only indispensable tool. Note the drawknife — most of the time it's just there to impress visitors, but it is very handy when making oars.

Mankind, the tool user — philosophers use this definition to distinguish us from the apes (although there are times when I wonder), but it could be that the humble craftsman, engaged in the creative use of mind, hand and tool, is exercising the highest functions of reason and logic.

Not only does the use of mind and hand together create a unique asset — something of inestimable value to its creator for it contains something of the builder's soul — but the gentle work helps unwind the tensions generated by living in a world overfull of politicians, financiers and insurance salesmen.

The tools needed to free this inner creativity are not beyond the ordinary person. In this chapter I will discuss the tools I use to build light plywood boats — firstly those which I would not be without, then those which I find useful to have.

My tools — indeed all woodworking tools — fit into the following categories. I will go through each in turn, concluding with some general notes on power tools.

1. Hitting
2. Sawing
3. Planing
4. Cutting
5. Drilling/Boring
6. Fastening
7. Abrading
8. Sharpening

The hand-held power jigsaw is a primary tool for building small craft. It is much easier to take the saw to the work than vice versa.

Hitting

Whether driving nails and punches, knocking in bolts, or just generally persuading things, the hammer is one of the primary tools, and I use three. The essential one is a light (400 g) claw hammer. The big 'clubs' that the chippies use to put a 100 mm galv spike in with three hits are not ideal for tapping in the small nails used for our little boats.

I also have a tiny upholstery tack hammer, handy for nailing along the edge of a sheet where it's up against the side of the boat, or other difficult work.

Number three is the 'ultimate authority'; a 1 kg brickie's hammer. This is my 'persuader', my 'dolly' for holding against unsupported timber when nailing, an anvil when held in the vice, or even a weight to hold two items together while the glue goes off.

I wouldn't be without the 400 g claw hammer — although an ordinary carpenter's claw hammer would do — and a good nail punch. Sandpaper the face of the hammer a little to reduce slippage, and go easy with the force of the swing.

Sawing

Here we begin to look into the Aladdin's cave of power tools. Without care the impulsive buyer could end up with a lot of shiny new toys and no cash for materials, so take care.

To break down the functions, first of all we need a means of cutting curves in up to 25 mm wood or plywood. Possibly the most used tool in my workshop, certainly the power tool that does the most mileage, is my Bosch variable-speed pendulum-action jigsaw. It does the job of a bandsaw, a circular saw, a crosscut handsaw, a rip handsaw, a power hacksaw, a power file, a circle cutter and has a myriad of other functions. Specialised machinery might perform each function better, but as a multi-purpose tool the jigsaw, with its many different blades, is my pet.

Next is a 10-point (per inch) panel saw; mine is a professional-quality Sandvik. It cuts quickly and easily with little breakout in plywood. It holds its edge well, something that many cheaper saws don't do in plywood. It also feels good to use.

When buying a handsaw, it is often worth searching the second-hand tool shops for old saws; Slack Sellars, Jack, Spear and Jackson, Disston, Stanley, and Kinzo are usually trustworthy names. Even if the teeth are pretty sad, take your treasure to the saw doctor and make it clear that it should have a fine set for finishing work. Before taking your selection to the counter, make sure: (1) that the blade is straight — no bow or kink; (2) that there are no rust pits near the tooth edge; and (3) that the handle is complete, and solidly fastened.

When buying a new handsaw I suggest you steer clear of the special 'hardened tooth' models. You will eventually hit a piece of metal with it, and if your saw doctor does sharpen it for you (which is unlikely!) he'll charge you heaps for ruining several files.

I also have two keyhole saws, Japanese cabinetmaker's saws (not easily come by, and I had to learn to sharpen them myself, but the fine cut and accuracy are worth it), tenon saws (two, hardly ever used), circle saws (hole saws, more properly in the drilling/boring section), a lightweight 18 cm power circular saw (commonly known as a Skilsaw™ except mine is a Black and Decker), several handsaws of various tooth spacings from 14 point to 6 point, and a saw-bench/buzzer combination.

The little Black and Decker circular saw was chosen for its light weight. I use it for heavy work that I know it wasn't designed for, and for a lot of ripping (stringers etc.). It had a set of bearings and brushes a while back, but otherwise has survived ten years of constant abuse.

My sawbench is a Cambro 20 x 10 cm combination buzzer and sawbench; it was cheap at the time, and the less said about it the better. I got caught with this one, and the only reason I still have it is I've not been able to unload it far enough from home to ensure the new owner doesn't bring it back!

My 'wouldn't be withouts' are my Bosch jigsaw, Sandvik 10-point panel saw, and a very fine

A selection of saws — bandsaw, jigsaw, Black and Decker hand-held circular saw, Japanese cabinetmaker's saw (Dozuki Ryo — cuts on the pull with a very smooth fast action, magical once mastered), and a fine 10-point panel saw. I have more, but again, the essentials would be just the jigsaw and the panel saw.

In the planes category: wooden-bodied Primus jointer; Jack; Stanley No. 4 & No. 60½ low-angle plane; Stanley rebate plane (bull nose); roundmouth plane (home-made); three miniature spokeshaves — more useful than you'd think; roundmouth spokeshave; straight wooden-bodied shave; Stanley spokeshave; Ryobi power plane. While I use all of the above, the two Stanley planes and the spokeshave are on my 'essentials' list.

tenon or dovetail saw. If you're shopping for a jigsaw, go for the top-of-the-line variable speed, variable pendulum action, with all the extras. The next addition would be a hand-held circular saw with tilt, rise and fall, and a rip fence. Don't get a great big one; the work is light, and the smaller unit is much easier to use.

Bench Saws

My preference would be for a 20 cm rise and fall and tilt. The old heavyweights are the best. A lightweight sawbench is about as much use as a lightweight steamroller! Tanners or Dycos are often available second-hand (Tanners are still

made in Penrose). A combination saw-buzzer is handy in a workshop short on space but we're getting into luxury now; bandsaws are also wonderful things but definitely in the luxury league.

Planing

Although I have a buzzer, a thicknesser and a couple of power planes, my hand planes are my 'wouldn't be withouts'. A Stanley 'Bailey' number 4 bench plane is the first item, and for a long time was the only one I owned. At 245 mm long and 50 mm wide in the blade it is a first class general-purpose tool — definitely a necessity for smoothing, straightening and trimming.

Next is a little low-angle block plane, good for trimming and rounding off edges, end grain and the like, rather than shooting a straight edge. This little plane, only 150 mm long, is usually used in one hand, and is a particular favourite of mine.

Boatbuilding is a series of small jobs — well suited to hand tools, but power tools, such as the power plane, are the real timesavers.

I have quite a selection of other planes, ranging from metre-long 'shooters' through various wooden and metal bodies down to a tiny 40 mm long violin maker's plane which is cute, but not very useful. Being realistic, apart from the two mentioned above, the only one I would suggest to add to your collection is a rebate plane; mine is wooden bodied, but Record, Stanley and Footprint all make good metal-bodied rebate planes, often found in second-hand tool shops.

Also in the 'planes' category are spokeshaves. Again, of my many examples, only one gets constant use. My pet is a 60 mm wide wooden spokeshave — light, easy to use, and holds a good edge — but these are less common; again, Stanley, Footprint and Record make good ones. You need the 'shave' with the rounded rather than the flat sole, and — like other second-hand tools — look for blades with no rust pits.

My power plane gets quite a lot of use. I have an old Ryobi, and I use this for really rough stuff, including planing lead keel castings! My number one unit is a Bosch, with reversible tungsten carbide blades. It has a rebate facility, a rip fence and a dust bag. The dust bag is a real boon, as the mess these things make is unbelievable. Note that these are scary tools to use, and need real care on the part of the operator.

Be careful with power planes; I would rate them as one of the most dangerous tools in the shop. However, for certain work they have no equal.

Buzzers

These are useful, but if you're on a budget, a power plane upside down in the vice would work — or just the longest plane you have and some elbow grease. Remember, you can achieve anything with hand tools; they just take application and time. Don't put a project off because you've no power tools — get started, and your hand tools will become your friends.

Thicknessers

Yes, I've got one. It enables me to produce usable timber out of demolition, waste or rough-sawn wood. However, bear in mind that most designers specify standard timber thicknesses, and your local timberyard carries these already dressed. You need to be doing an awful lot of work to justify such a major piece of woodworking plant as this.

Cutting

This category includes chisels, knives and so on, and the choice is simple. A 12 mm straight chisel, a 38 mm straight chisel, and a Stanley knife are my 'musts'. I've lots of gouges and specialised carving chisels, but the two above and the Stanley knife do almost anything. The bigger chisel, in particular, is often in my hand. It is, for me, a small plane, carving tool, chisel, bevelling tool and pencil sharpener.

A note on chisels in general; I have one fancy wood-carving mallet which only gets used when I'm doing some fancy wood carving; otherwise, I generally hit my chisels with a hammer! The plastic-handled 'carpenter's' chisels (Stanley again) that I favour for general use will stand any amount of this, but remember, if you have to swing your arm rather than your wrist to drive the chisel then you're probably not doing the job the right way, you are using the wrong tools, or they are blunt.

Dad used to make wood-carving knives for me from industrial hacksaw blades, rubber tubing and masking tape. The Stanley knife is good but for some reason the blades don't sharpen easily, so you've got to keep a packet of fresh ones on hand. If you can get a blunt industrial-weight hacksaw blade from an engineering works you might like to try your hand at making a knife. Sixty millimetres is plenty of length for the blade. The edge should be dead straight, and the handle a little longer than the width of your palm. Don't overheat the steel when you're grinding the blade to shape, and finish up on a fine oilstone.

Drilling/Boring

For what I do, a small hand drill (egg-beater type), with a 6 mm chuck and a brace and bit are quite adequate. The egg-beater uses engineering twist drills, and — apart from the few occasions when long or difficult holes have to be drilled — is almost as quick as a power drill. You might say it's a 'cordless' drill which never has to be recharged.

The brace uses auger bits, and using them I've drilled 30 mm holes 200 mm into hardwood.

My selection of things I used to use to make holes: an old veteran two-speed Hitachi 9.5 mm drill, a couple of egg-beaters, one of which is a two-speed 9.5 mm Record, one of four brace and bits I own, and my 'universal' wood-worker in drill press mode; in spite of its many functions, this machine is only rarely used. I have, since this photo was taken, inveigled Mum into making me a gift of a 9.6 volt Makita cordless drill which has made almost every-thing else redundant.

The bits can be bought quite cheaply second-hand. The brace can also turn a screwdriver bit that drives big screws very effectively. I have three of these: one for drilling holes, one for the screwdriver bit, and one for a countersinker.

Power Drills

The danger in getting a great heavy piece of equipment with all sorts of features and heaps of power is the thing is so clumsy that it's a liabili-ty for small work. I use a 10 mm chuck reversible two ratio variable speed which is real-ly great for big heavy work; I've cut a 50 mm hole in hardwood with an adjustable trammel

bit and had power to spare. The slow speed facility is particularly important for drilling stainless steel.

I'd been looking for a lightweight 6 mm chuck drill for some time, and recently obtained a cordless driver/drill which is a real revelation in timesaving. It drives screws, drills pilot holes, and is so good I'm going to get another one so I can have one for each function!

A drill press stand for the power drill, or a proper drill press, is a real asset, as much of our metalwork is stainless steel or bronze — but again this is on the luxury list.

Fastening

Screwdrivers are a subject that often gets over-looked. In addition to the brace and screwdriver bits I have several wooden-handled cabinet-maker's screwdrivers (Marples still make them). The engineering screwdrivers are much too small in the handle, hard to turn (no leverage), and tough on the palm of the hand when push-ing hard. Screwdrivers should be sharp-ened as soon as the edge gets rounded so they stay in the slot.

There are lots of different screw heads about. Philips (cross head) and Posidrive (modified cross head), are not among my favourites; they go well with power drivers but I find them not to my liking for hand use. However, I've just dis-covered Scrulox square drive screws. Scrulox make screwdrivers in many shapes and sizes, driver bits for drills, cordless drivers, and even brace and bits. The screws are available in stain-less, and I've found them so much easier that I'm in the process of changing over from slotted screws. Scrulox hold a large proportion of their home Canadian market, and having tried them, I can see why.

Nails are my primary fastener, mostly small holdfast (ringed), flat-head silicone bronze or monel nails. Square copper nails with rooves, as used in traditional clinker building, are 'in stock' in my workshop but are rarely used. Galvanised nails are better in theory than in practice; it's hard to drive and punch them with-out disturbing the zinc coating, and they are now used only in jig building. I did a couple of construction ply quick-and-dirty skiffs a while ago, and in spite of punching the galv nails in well and plugging with epoxy filler, they were bleeding rust all over the white enamel within twelve months.

The ring barb (holdfast) nails hold well in most timbers, and a pair of them at different angles can almost equal a good-sized screw in holding power. If you're using these, be careful where you put them as they can be hard to pull out — after all, that's why they're 'barbed'. If extraction is necessary you may need to break the head off and punch it right through with a fine pin punch.

Temporary Fastenings

Clamps are a whole subject on their own, and often a problem for amateur builders. My collection, built up over quite a few years, is worth more than I'd be prepared to spend in one hit. Only someone contemplating a long-term career in boatbuilding should contemplate this sort of investment. However, the top-of-the-line woodworking clamps — the sliding bar, or F-clamps — are fast in action, have fair 'pressure' and a wide 'mouth'. They are my preference but usually need two hands to operate, sometimes a disadvantage.

Engineers' G-clamps, often cheaper than F-clamps, have a greater pressure and a reasonably wide mouth, but still require two hands to use in most cases.

Spring clamps (like spring clothes-pegs) are a viable option. I have heaps of them; if one won't hold I use two or three or more. Their one-hand operation offsets the low pressure and narrow mouth. They're cheap, and — if you're careful — don't noticeably mark the work (I've actually used spring clothes-pegs on small work). It is easy to make wooden 'nipper' clamps with scrap wood, a piece of leather and a bolt.

I also hold work together with pan-head PK (self-tapping) screws with pads under the head. A 25 mm square piece of ply with a 5 mm hole through it makes a good pad. Note that the hole in the pad should be a slightly loose fit on the screw. On the same subject, if you wish to draw two surfaces together with a screw, it pays to drill an oversize pilot hole for the piece under the head of the screw, otherwise it will be very difficult to achieve your aim once the thread of the screw is engaged in the wood. The holes from the screws can be filled with epoxy fillers without loss of strength, although the fillers tend to mar a varnished finish.

PK screws can be effective in areas where a clamp can't reach, and are particularly useful when clamps are in short supply. My rule with clamps is that whenever I see a special offer I buy one or two — if you're going to make a hobby of boatbuilding they're a real asset.

Abrading

This includes sanding, grinding, and filing or rasping.

Sanding

Sandpaper is something that most of us don't consider, except that it doesn't last long and costs a lot for such a mundane item. It is so expensive that it is worth a little research.

Take the sand part: much of the grit on do-it-yourself paper is garnet or crushed stones — hard stone, but not very strong, and as a consequence doesn't keep the sharp points that make it 'cut' for long. Its only virtue is that it's relatively cheap.

For our use, Alox (aluminium oxide), blue or green, is both readily obtainable and fairly effective. If you have a choice, buy 'open coat'. These papers tend not to clog up as much as 'standard coat'. I have several grit sizes on hand; 40 grit for rough shaping or quick work, 80 grit for smooth shaping or intermediate finishing, and 150 grit for fine work. I've many other grits, such as 26 grit for the disk sander, and 'wet and dries' down to 600 grit for paint work, but 40, 80 and 150 grit covers most woodworking applications — unless you are a perfectionist, of course.

Those who are in a position to do so should phone an abrasive supplier such as Morris Black and Mathieson, or Carborundum. These people make wide sanding belts (sometimes 8–10 square metres in area) for commercial machinery. The offcuts from the bulk rolls of material are often available at a good price. Bear in mind that the stuff is of a quality that will make the hardware store stuff look like a waste of time. It's not cheap — but it will last longer. I have some 2 metre wide pieces of Ecamant 80 grit black open-coat cloth back, and I go to the trouble of cleaning it with a soft wire brush. It's amazing stuff; I reckon it will last me for ever!

Buy the best you can get — it's cheaper in the long run.

Some sandpaper is advertised as being 'non-clog' having been treated with zinc stearate to reduce the adherence of paint and so on to the abrasive material. Some paints and varnishes do not like zinc stearate, and it will 'crater' and peel if applied to a surface prepared with this type of paper.

With regard to sanding blocks, cork blocks from your DIY store are very useful. The plastic ones with clips on are not, in my experience, so good. Most of my sanding blocks are just offcuts of wood or dowel with the edge shaped — effective and cheap. Make your blocks 200 or 250 mm long, and keep moving the paper around it to keep it 'fresh'. Use a smooth steady action and you will be surprised how effective elbow grease can be.

Note that the 'pros' use things called 'longboards', flexible boards that are usually around 100 mm wide, with a handle at each end, ranging between 600 mm and 2 metres long (definitely two-man jobs). These only use sandpaper on a short section 'midships', the idea being that a longboard only knocks the tops off the bumps and — being flexible — will follow around a round-bilge hull to produce a finish with no 'unfairness' in it. Longboarding a big hull is a job that causes boatbuilders' apprentices to catch the flu — at least that's the impression their workmates get as they're so often AWOL on longboarding days.

Grinding

I have a little 100 mm Bosch angle grinder; it's not on the 'can't do without' list but it sure gets a lot of use. A boatbuilder friend of mine calls his the 'power spokeshave' and that gives a good indication of both the uses and capabilities of the thing. In fact, sanding/grinding is the area of boatbuilding that is the easiest to do entirely by hand (requiring the least skills) but in which the biggest time savings can be made with the appropriate machinery.

Sanders come in several configurations; I will briefly run through them; from the least to most useful to me.

Disc sander — as a finishing tool forget it. Mine is only used for carving wood, and then mostly for shaping oar blades.

Belt sander — some of the little lightweight ones are not too bad, although the edges of the belt do tend to leave grooves. If you have got one, use it with care, but I would not go out and buy one. I mainly use mine upside down in the vice for pre-finishing components.

Orbital sanders — the cheaper ones tend to be an example of getting what you pay for. I use two particular orbitals a lot; a large 'half sheet' size Bosch, expensive but an amazingly effective machine; and a little Makita palm sander which gets used a great deal for light sanding between coats of paint — particularly on boat interiors.

The other device I use is a PEX 250 sander (another Bosch — I find their tools consistently good, both in reliability and user comfort) where the machine combines a rotary and orbital action. This thing is great, and if I had come across it first it would have saved me buying about three other sanders.

If you're going to buy a sander, I would recommend a really grunty commercial-weight orbital, or the PEX rotating/orbital or eccentric sander however, read my final note on power tools first.

Files and Rasps

My own pets are coarse single-cut smooth-edged engineers' files and a similar large 'rat tail' file. The coarse woodworking files have no place in my workshop. One oddity I do find useful is a large flat panelbeater's 'body file'. Its large semicircular teeth and slightly flexible construction make it ideal for trimming and smoothing the epoxy and fibreglass I use so extensively.

I have lots of others but the above items are really all that get used for actually building the boat; the others are for either making fittings or sharpening the tools.

Sharpening

Sharpening your own tools is a must. Forget saws (both hand and circular); you can expect to know a little in advance when one needs a touch-up, and the local saw doctor will do a good job for surprisingly little money. It is the chisels and planes that need regular attention.

You could build this little boat, or any of the smaller boats in the design section, with only the tools in the box: jigsaw, fine panel handsaw, hand drill (egg-beater), drill bits, clamps, measuring tape, pencil, screwdrivers x two, chisels x two, spokeshave, hammer, countersinker, sandpaper and a plane (Stanley No. 4).

I was lucky enough to find a pair of Japanese waterstones in a second-hand tool place, even luckier in that the proprietors didn't know what a treasure they had and were asking only a couple of dollars for the pair! These are clean and fast cutting, and with care will give an unbelievable edge. I have 800 and 6000 grit stones; keep them in a tray of water and go to the trouble of wiping the tool off with oil after sharpening. If you can find such stones I recommend them, but there is a considerable difference in technique required, so be careful until you are familiar with them.

For years I used a double-sided Norton (coarse and fine) synthetic stone and I still recommend them for most people. The coarse is quick enough in action to grind out the nick you made when you hit the head of a nail, and the fine will give you a quite acceptable edge.

A power grinder is useful, and the double-ended bench grinder is now cheap enough to warrant one in most workshops. Do change one of the wheels for a soft fine one suited to grinding cutting edges, and keep a tin of water alongside to cool the work off to prevent blueing. The machine itself should have an adjustable tool rest and a plastic screen. However, even with this screen down you will need safety glasses.

I have a few odd triangular and flat fine-files which I use for auger bits, and other odd cutting instruments, but my stones are my primary sharpening system. Look after them and they will be your friends for a long time.

Notes on Power Tools

Only buy the absolute minimum! It is very easy to get carried away. I have 26 power tools of various sizes and shapes, and you will note that in the above text I haven't even mentioned things like routers (I have three), which many find indispensable. Of all the tools I have though, I could get away with only four or five. Think hard and buy a couple of the very best items you can afford rather than a lot of cheapies. Buy a brand which can be locally serviced and is in common use in your area.

When you start your second or third boat you will have an idea of just what will improve your work rate for future projects.

Chapter Five

Basic Joining Techniques

Scarfing and Joining Plywood

Of all the operations that face the beginner wishing to build a small plywood boat — next to lofting, or full-sized drawing of the vessel before starting the 'real' work — scarfing ply sheets end to end creates the most apprehension.

Scarf Joints

A scarf joint is a bit intimidating if you've never seen it done before. A dead-straight long taper the width of a sheet of ply, it requires a standard of accuracy exceeding even that demanded of us by our school woodwork teacher. Mine used to wander the shop muttering 'Cut on the waste-wood side lad, cut on the wastewood side.'

However, the old gent didn't have the advantage of the enormously strong, gap-filling epoxy glues we now have, and as a craftsman of many years' standing was often as concerned with using the correct method as with the result. We, on the other hand, should be concerned mostly with the result, and need to use whatever method yields a satisfactory joint, one that is easy to create and yet will not fail under the most extreme load contemplated — and that with a large safety margin — I don't like swimming.

Scarfing plywood, often called for in today's lightweight boats and necessitated by the limited size of the ply sheet, is nowhere near the struggle you might imagine. I was shown the system outlined below by a boatbuilder far more skilled than I but even more impatient. Time was this guy's enemy, and he got more done in an hour than anyone I've ever seen. The laborious method of cutting each bevel and carefully matching them with chalk that old 'cut on the wastewood side' taught me would have driven this man to violence or fibreglass.

His method is simple, fast and effective, demands no special tools (but is definitely a lot faster using a power plane), and with care will give a consistent result.

Step 1: Decide on Your Scarf Slope
I use a 6/1 slope unless it's a particularly highly stressed joint, then I use the 8/1 slope that the West System people suggest in their book. I've used 4/1 but can inform you with authority that it is not as good as a longer scarf might be (I nearly got wet).

These ratios are the distance along the joint as a multiple of the thickness of the ply being joined; that is, 6 mm ply = 6 mm x 6 = 36 mm; six times as long as it is thick = 6/1 (Figure 1).

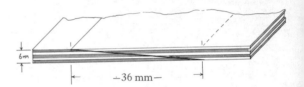

Figure 1: A 6/1 scarf.

Step 2: Cutting the Bevelled Edge
When building the boat it is often of assistance to plan the work so you'll be able to make the major scarfs before doing anything else; that is, before cluttering an often limited workspace up with building jig and frames.

On my 6.4 metre trailer yacht we need a bottom panel 2½ times the length of a standard 2400 mm long sheet of ply, three cabin top panels 2800 mm long, and two cabin sides/cockpit coamings 3500 mm long. The bottom panel and cabin top panels are cut from 9 mm ply, and the cabin sides/cockpit coamings from 7.5 mm ply. With a little cunning and forethought these can all be scarfed in one operation.

Having worked out where on the sheet the components lie, and cut the sheets to length allowing ample extra for the length lost when cutting the scarf, we have a pile of sheets, in pairs, in two different thicknesses.

A scarf join in the process of being planed — in this case, two pieces of 6 mm plywood.

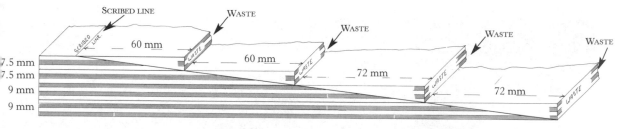

Figure 2: Laying out the scarf slope in two sheets of 7.5 mm ply and two of 9 mm ply, at 8/1.

First, your scarf slope; some of the joints in this case are very important — they keep the water out — but are quite well supported, so I specify an 8/1 scarf slope. Nine millimetre ply means 9 mm thick and 72 mm long; 7.5 mm ply means 7.5 mm thick and 60 mm long in the scarf. Do measure the thickness of your ply before you calculate the length of the joint; a nominal thickness (the thickness it is meant to be) may bear only a slight resemblance to what you actually get, and this is one thing that should be right. For the purpose of this exercise we'll assume the manufacturer has got the thickness of your plywood right.

For the 9 mm ply, take a mortise gauge, or make one from a scrap of ply, a nail and a pencil, or use your adjustable square the way old 'cut on the wastewood side' taught you; and scribe a line 72 mm from the end. This line should be as accurate and as straight as you can make it. For the 7.5 mm, do the same 60 mm in from the end of the sheet. Lay out some 6 x 2s, or similar, to support your panels full length off the floor; you won't want to hit the concrete with your power plane. Take care to see that there is no sag, particularly across the end being joined.

Stack the ply, marked side up, with the marks all at the same end. Very carefully slide the sheets until all of the ends line up with the scribed mark on the one below it and with the sides of the sheets in line. You'll end up with a series of steps (Figure 2).

Check that the stack cannot sag and is solidly supported where you are going to work. Clamp the stack well back from the scene of the impending action so the clamps are out of the way. If in doubt, nail right through just back from the mark on the top sheet; you can pull the nails out later, and fine panel pins don't leave much of a mark. There must be no possibility of any movement during the energetic exercise to come.

Step 3: Planing the Scarf Slope

Your job is to achieve an even slope from the scribed line on the top sheet to the lower end edge of the bottom sheet (Figure 3). Being made of parallel layers of wood and glue, plywood, when planed straight, exhibits a face of straight

Figure 3: Scarf slope dressed straight.

and parallel lines that resemble stripes. If planed crookedly the lines appear to have humps and hollows; this provides a good guide with which to get your scarf slope accurate. Keep going until the steps merge into the smooth slope; the bottom edge becomes a sharp or 'feather' edge, and the planed slope is cut back to the scribed mark across the top sheet. Work away carefully until the slope is dead straight in all directions. Although it is no problem to do the whole job with a hand plane, roughing out the slope with a power plane makes a laborious job a real breeze.

Having done the hard work, glueing up is a piece of cake. Lay the first pair of sheets out with the slopes overlapping (Figure 4) so there is neither a bump nor a gap. Accuracy is very important here and can be checked by running a string line along the edge of the pair of sheets; once the straight line has been established the others can simply be lined up on the pair below.

If you have a problem lining up the joint on the slippery glue, nail the ply through into the bottom 6 x 2 so it can't slide out of alignment. Use a piece of string to line the sides of the sheet

A 'zigzag scarf' joint. This is the bottom of a Rifleman *runabout.*

Figure 4: Scarf in position.

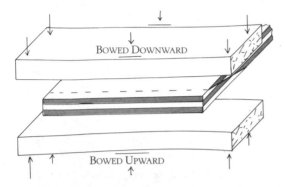

Figure 5: Clamping.

up, and check the joint carefully before fixing in place. For your stack of sheets just keep laying up pairs with plastic wrap in between the glued areas, nailing off with fine panel pins, up to four or, if you're using if thin ply, maybe six layers maximum.

With one of the 6 x 2s directly underneath the joint area, put the top 6 x 2 on over the top (Figure 5). Put a pair of heavy screw clamps on each end of your pair of 'pressure bars' and wind them down. You are using the bows in the heavy wood to generate clamping pressure evenly across the width of the plywood sheets.

Clamping the ends too hard can lift the pressure off the centre — try and get the clamps in over the edge of the plywood. At all times keep the sheets lifted off the floor flat and straight; any sag or unevenness will produce a glued-in bend.

Leave this assembly for at least 24 hours, in cooler weather perhaps twice that. These are joints that may be asked to take a high loading, and it's better to give the epoxy some time to cure to near-full strength.

When your clamp has been disassembled you should find a sound joint with a little excess glue to be planed off with a fine-set hand plane (an instrument I find better than sandpaper for this job), leaving a long sheet of ply very near as

strong as a single unbroken sheet.

A quick-and-dirty method of achieving a similar result, not really any quicker, no dirtier, but not requiring the accuracy of the real thing, is the taper and tape splice.

Bevel the ends to be joined so you have a shallow vee with about a 6–8/1 slope; accuracy is no problem. (I've seen a bevel produced with an angle grinder — and I didn't like the rest of that boat either!) Note that the bevels are on the same side, that is, on the top face.

Lay your ply out, check the position, and secure in place. Weights are okay as there is not the same possibility of movement (plastic wrap underneath again). Wet the joint area with epoxy resin, then tape up with glass tape (Figure 6) until the vee is slightly overfull of wet-out glass cloth.

Figure 6: Taped splice.

Put a layer of plastic wrap on top, then put a flat plank across with weight on it, and wait for a day or two as before. When you are happy with the rigidity of the joint you can mow the excess off flush with the ply surface, again with a plane.

There is another method, particularly suited to end-joining thin plywood. I call this the 'zigzag' scarf and use a 30° degree set square to mark out the zigzag on a 60 mm spacing. After cutting out the first sheet, I overlap the sheets and use one as a template to mark out the other. When cut out with a fine-set saw or jigsaw, it is a simple glue job to butt join them. The same care is needed to keep the sheets parallel and straight. This join is simple to make, easy to get 'fair' in a thin material, and seems strong enough to cope with normal loads. I have used it in a number of applications and am happy with the results so far. My thanks to my friend Julian Godwin for the idea.

When ply boats were built in the old way, they had a great deal more framing than is often the case today. The eight-foot sheets were applied

Figure 7: Butt strap with fastenings.

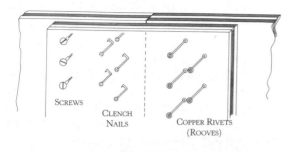

SCREWS

CLENCH
NAILS

COPPER RIVETS
(ROOVES)

Figure 8: Epoxy fillet tee joint.

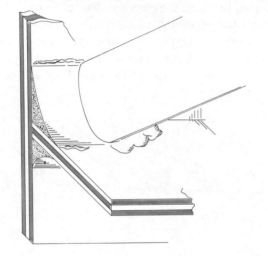

8A

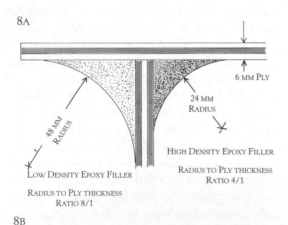

6 MM PLY

24 MM
RADIUS

48 MM
RADIUS

HIGH DENSITY EPOXY FILLER

RADIUS TO PLY THICKNESS
RATIO 4/1

LOW DENSITY EPOXY FILLER

RADIUS TO PLY THICKNESS
RATIO 8/1

8B

one at a time, and were joined using 'butt blocks', 'tingles' or 'butt straps', all the same thing, a method of beefing up a straight butt joint in wood by fastening another piece of material on the back of the joint (Figure 7).

Fastenings vary from countersunk head screws on thick material, clench nails, rivets, or just plain epoxy glue. PK screws are a good way to

keep the pressure on until the glue has gone off, and then it's an easy job to remove the screws and fill the holes with epoxy filler.

Remember, butt straps should be of material no thinner than the pieces being joined; the grain direction of the butt strap's outer plies should be at right angles to the join line; and the width of the butt strap should be at least sixteen times the thickness of the ply being joined (8/1 each side).

If you opt for the butt straps, and I do in certain areas of the bigger boats, it is often possible to plan the joins so they fall in inconspicuous areas of the boat, such as against the frame (allowing for the width of the strap!), or behind a bulkhead. In addition to its main function of joining, a heavier butt strap can be used to reinforce or stiffen an area, such as the floor or bottom of a light dinghy where the occupant may stand on a thin bottom.

Other Joints
Tee Joints in Ply

There are two common methods of producing a tee joint in plywood — the epoxy fillet and the corner reinforcing block.

Figure 8 details the epoxy fillet with little further explanation necessary. Do use high-density filler if possible; its much higher compressive strength makes for a much stronger joint but one with much greater rigidity.

The corner reinforcing block can be just a simple block (Figure 9a), beefed up with screws or holdfast nails (Figure 9b), or perhaps, for a joint loaded in two directions, a glued-in block on both sides (Figure 9c). Variations in the size of the block occur according to load, direction of load, and structural consideration of the components being joined. As a rule, for 6 mm ply or less in thickness (or thinness, as the case may be), I use 20 mm square 'corners', not because of some esoteric engineering formulae but because it cuts easily from 25 mm dressed timber (finished 19 or 20 mm), and is as small as I can easily drive nails and screws into.

If light weight is of serious consideration I do get down to 20 x 12, but only on very lightly loaded areas do I go smaller. One project, a plywood touring kayak, has 9 mm square corners on the 3 mm coamings (underside), and a 20

'Trowelling' in an epoxy fillet prior to glass taping a seam in Seagull.

mm radius low-density fillet on top where the aesthetics are important.

For heavy joints, like the main bulkhead in my 8.5 m (28 ft) cruising cutter design, the 9 mm bulkheads have 20 x 60 on each side of the 9 mm bulkhead. This provides plenty of glue line to the ply (the 60 mm faces), and a good solid area for driving in the fastenings to hold the 18 mm strip planking in place while building and later the bulkhead fixed within the hull under rigging torsional loads.

It doesn't take a lot of time to greatly increase the strength of the joint. It's tensile strength that most tee joints are deficient in, so use a bigger fillet on the tension side, or beef up the blocks with screws or ring barbed nails as shown. Note how the fastenings overlap within the timber, reducing the chance of splitting. Remember that you'll need quite a few nails to equal a screw.

Ply Lap Joints

These come in three variations at my place, and they are the joints used in plywood clinker-style boats. Many of my boats use this system to produce compound curved hull forms without the more complex, labour intensive methods (cold moulding, carvel planking, strip planking, and so on), more suited to the bigger boats.

On a very light boat, a lap joint such as type A (Figure 10) can be appropriate; however, it is necessary to make dimension B six times dimension A. This is very hard to achieve in practice as the angles at which the pieces meet sometimes do not allow that slope (bevel) length — hence type B (Figure 11).

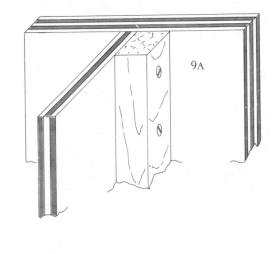

9A

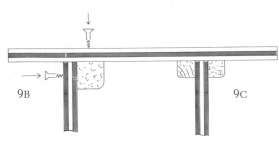

9B   9C

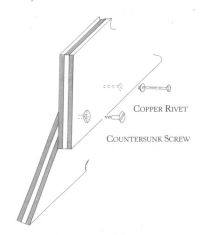

COPPER RIVET

COUNTERSUNK SCREW

Figure 9: Tee join with corner blocks.

Figure 10: Lap joint A.

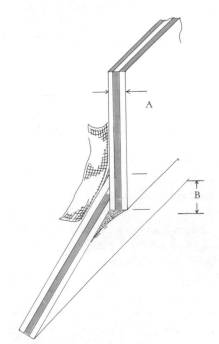

Figure 11: Lap joint B.

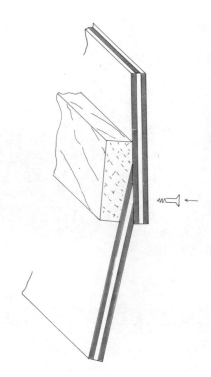

Figure 12: Lap joint with seam batten (stringer).

The epoxy fillet under the lap and the glass tape on the 'back' greatly increase the strength of the joint. These joints can be further increased in strength by using screws (in heavy plywood), clenched nails ('clencher built' is where 'clinker built' came from), or copper nails with rooves (conical washers). I've built boats up to 8 metres and 2200 kg displacement like this, but much above 6 metres the method that follows is easier for the average builder.

This more traditional method can be used with appropriately sized stringers on everything from tiny yacht tenders to very large craft. It is much easier to achieve a 'fair' line, and slightly lighter plywood planks can be used, but you'll lose the clean interior that the taped reinforced lap has (Figure 12).

Glass Taped Butt Joints

Stitch and tape, tack and tape, sew and stick — all (very slight) variations on a method pioneered in the UK by Jack Holt of Mirror dinghy fame. I've no doubt Jack was not the first to use this method, but he was certainly among the first to build a large volume of boats in this way.

The system uses the tensile strength of fibreglass and the compressive strength of the wood and fillers to create a joint that requires no other materials, that is stringers (Figure 13); remem-

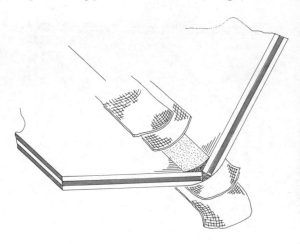

Figure 13: Epoxy-filled and fibreglass-taped butt joint.

ber the taper and tape splice? Note that gaps are not a worry; within reason quite large gaps, perhaps 4–6 mm, can be filled with HD filler, and may be stronger than a tight all-wood joint, in view of the increased area of glue in contact with the wood.

Covered over with tapes according to the designer's specs, these joints should always break outside the tapes rather than at the angle where the ply meets. (Hopefully it won't break anywhere!)

A whole range of joint types in Rogue. *Almost ready to plank, it's beginning to look like a boat!*

I use nylon fishing line (40 kg monofilament) to 'blanket stitch' panels together. Although it's not easy to tie the knots, I can tape over the lot and plane the bumps off later.

Buying fibreglass tape is a quick way of emptying your wallet. (I guess if you were really worried about that, you wouldn't be boatbuilding!) Buy the appropriate weight of woven cloth, and cut pieces of tape off at a 45° angle to the weave. This stops the tape falling to bits in a horrible mess when you hit it with the resin brush.

Building a lightweight plywood boat is essentially a variation of the joints above, and even a large project is just a series of small, simple operations taken one at a time. Note that the difference between plans for 'amateur' construction and those for the 'skilled' worker — in plywood anyway — is that the former will tell you what each step is and in which order you should take them. There are exceptions, however; if in doubt, write to the designer, outlining your skills, and ask!

Chapter Six

Plans

Reading between the lines ...

So far we've looked at the primary material for the average amateur (plywood), the adhesives used to fasten it, the tools used to shape it and the way you join it. Hopefully some of you will be about ready for the off!

So, before we build the boat, we select a design or designer, and — after some negotiations and payment — our hypothetical amateur sits down at the kitchen table with a set of PLANS!

To the first-time boatbuilder a set of plans resembles half a dozen sheets of odd-looking drawings with gibberish written all around them. How do you go about converting these into a gracefully curved vessel that will convey you safely across dark and troubled waters? Read on!

Most plans intended for the home boatbuilder do not require 'lofting'. This is a process where a boat is drawn out life-sized, and the dimensions of various parts 'picked up' as required. This is something the designer should do for you, even if it is from the original scale drawing, as I do it, or — as some industrious souls do it — from their own full-sized loftings.

A few plans come with full-sized patterns for 'key' items such as frames, bulkheads, floors and transoms ('thwartships' or 'across' members), and stem, spine and keel ('fore and aft' or 'along' members). This desirable method is practical only with smaller boats; the paper used for patterns is dimensionally unstable, altering size and shape to a surprising degree with changes in humidity. It is possible to use films, such as polyester or mylar, for patterns, but cost makes them uneconomical.

Without full-sized patterns, at least for our medium-sized and larger boats, the designer must convey to the builder the shape of each major component in such a way as to make accurate translation from drawing to reality easy.

Before we go any further, there is a skill that all builders need to know: how to produce a 'fair

curve' from a 'set of offsets'. See, we're off into the gibberish already — but don't worry, it's not hard in practice.

A fair curve is a nice sweeping curve with no sharp changes, bumps or flats in it. It is much easier to see than describe — particularly on a boat! (Figure 1.)

A 'table of offsets' is not a dessert at a six-year-old's birthday dinner! In its simplest form it is merely a straight line with measurements out at right angles at regular intervals (Figure 2).

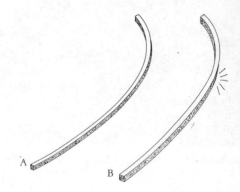

Figure 1: A. Fair curve;   B. Unfair curve.

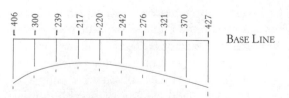

Figure 2: Simple offsets.

These tables do get very complicated, but like most complex operations, they are a series of simple but interrelated processes. The illustrated table is a set drawn solely as an example. A slice through the boat (from side to side) is taken; the measurements are from the centreline out, and the waterline up and down. The table shows how it works (Figure 3).

3A

On 200 mm Grid	
CL	-282
1	-256
2	-208
3	-108
4	+72
5	+---
6	
Gunwale	774
Gunwale	1094
5	---
4	---
3	980
2	953
1	878
WL	735
A	424
B	

HEIGHTS STAT 4

½ WIDTHS STAT 4

3B

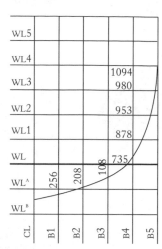

Figure 3: Metric table of offsets and application.

Richard Young marking out the sides for his Fish Hook.

Figure 4: Stem offsets from baseline.

Given this line of points on a grid, you need to draw a 'fair curve' through these to produce the shape required. For a stem, measurements are often taken from a pair of base lines, as shown (Figure 4).

To draw these fair curves, you'll first need to find a method of providing a stop or stationary point at each measurement point, then a long flexible 'batten' to bend around them. Two means of holding your batten are weights or thin nails. I've used tins of paint, bricks, scrap iron, and lead weights to hold the batten; almost anything dense and of sufficient weight will do while you first sight along and then draw along it. The 'pros' use fancy lead weights with wire points, but apart from being easier to draw

Converting the offsets on the plans into a graceful full-sized curve. I'm using a piece of plastic extrusion held by fine nails, a large square, a small square, and a steel ruler. Note the string line to show a true straight line for the 'base'.

around they don't really justify their expense for a one-off project.

The batten used will ideally be one of a selection with varying cross-sections and lengths to produce the different curves required. In practice some of the plastic extrusions used to join wall panels (available at your local hardware store) will do the tighter curves, and will bend very easily if used with care, while the longer curves can often be handled by the best of the stringers you've cut to build with. After all, it will have to follow the curve when included in the structure. Choose one that takes the curve without force, and preferably one that is at least twenty percent longer than the curved line required.

It may be necessary to cut one or two special pieces of wood, perhaps tapering a little at the ends to follow the curves easily. These should be from the straightest, most stable wood obtain-

able, and they need to be stored very carefully. Look after your battens; if you get a kink in one the kink may appear in your boat.

Wrap your batten around your weights or nails, following through the points you've measured and marked. You may have to slightly move a nail or two; any more and you should check and rethink. A fair curve is the important thing, more so than the nail that may be a couple of millimetres out. Always remember that not only are designers mere mortals, they are often working from a scale drawing, and tiny errors are greatly magnified when scaled up.

Remember to take your batten well past the points at each extremity, continuing the curve so you don't end up with a 'flat spot' at each end!

Hard-chine and multi-chine boats work a little differently, as the transverse shapes you wish to produce are often made of interconnecting straight lines. All that is required to draw them are the co-ordinates of the intersections. These points are usually measured out from the centreline and up or down from the waterline or a baseline (Figure 5). You will find that a lot of my small boats use this system.

When drawing your boat out, it is no problem to draw components one on top of the other. I buy a 'coverboard' (the damaged piece at the top or bottom of a stack) of particleboard or hardboard — a nice new sheet! I paint it with flat paint (isn't all paint flat?) to give a good surface on which to draw. For the cross-sections it is necessary to produce only half-expansions, so a 2400 x 1200 mm sheet will do for quite a large boat. I use an ordinary HB pencil (keep it sharp to keep the line thin) and mark the centreline and waterline permanently.

Having drawn your frames out full-sized, the next problem is to transfer the shape to the wood from which you will cut your components. A traditional method, and one that works well for me, is to lay a line of small flathead nails around the curve, points inwards and heads on the line (Figure 6).

Lay your wood carefully on top and tap it down sharply with a mallet; if you've spaced your nail heads right (closer on tight curves), it is easy to redraw the line on the 'real piece'.

Other methods include tracing onto heavy transparent paper and cutting out, drilling fine

The end result of the conversions of a set of plans into the full-sized item — a brand new Tender Behind.

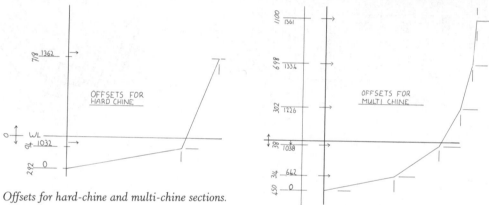

Figure 5: Offsets for hard-chine and multi-chine sections.

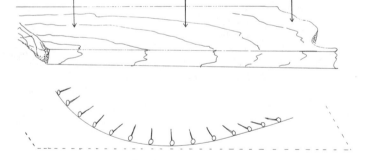

Figure 6: 'Picking up' with flathead nails.

holes through the drawing and 'pricking' the pattern through, using a dressmaker's serrated wheel to achieve a similar result, or even cutting the pattern out of a sheet of bisonboard. This latter method is particularly good for items one needs several of!

Now that you can draw a 'fair curve' from a 'table of offsets', 'loft' a frame or stem, and con-vert a stack of drawings into full-sized boat parts, we'll carry on in the next chapter with a look at the first of several methods of building small craft from plywood. I will cover 'tortured' plywood, 'stitch and tape', conventional 'frame and stringer' construction, and have a look at my own pet system which combines the easy part of all three methods.

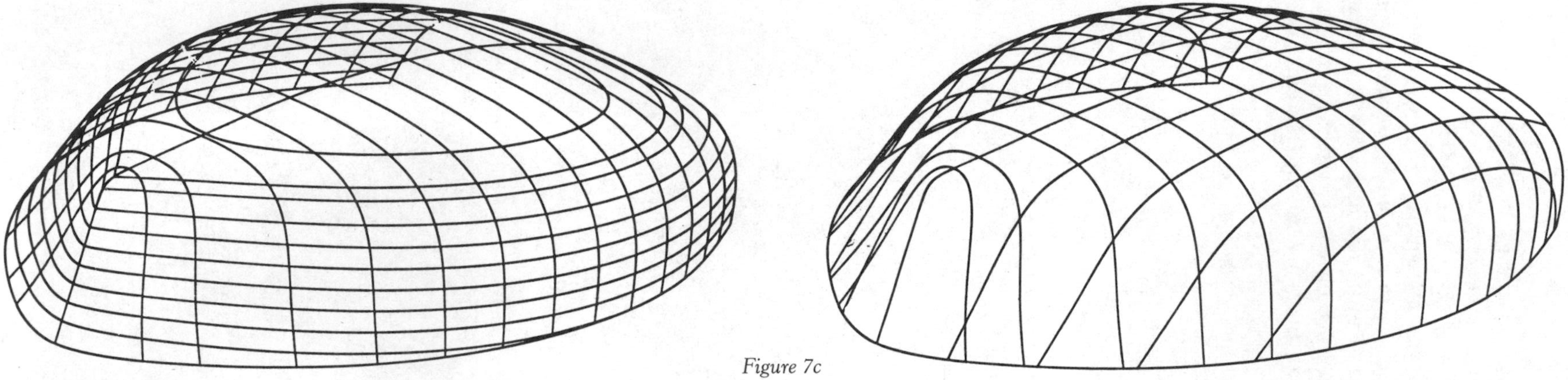

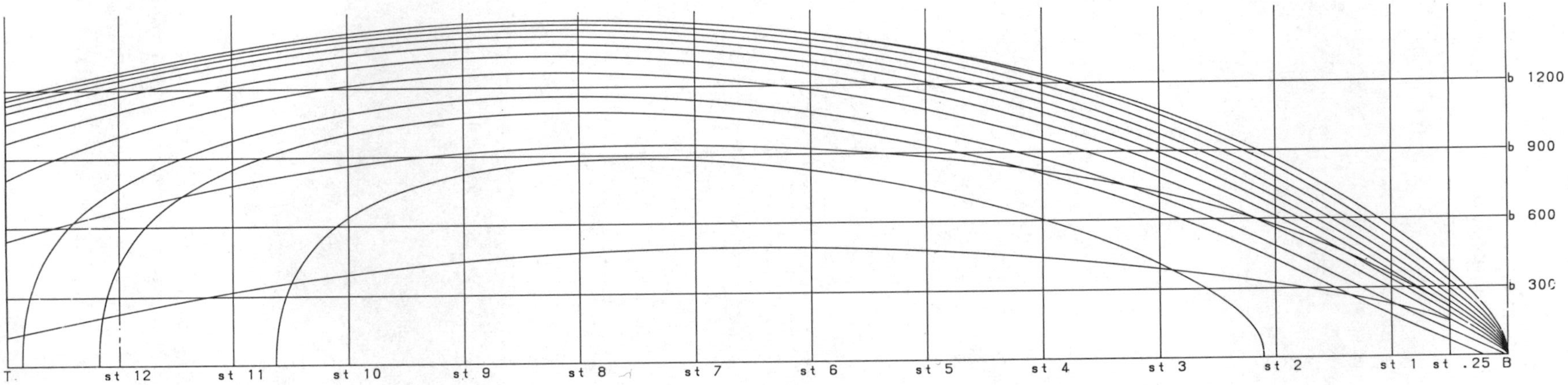

Figure 7d: A computer-generated line drawing for Essentially, including a pair of perspective views.

Figure 7c

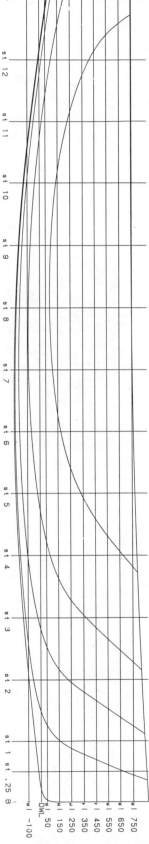

Figure 7a

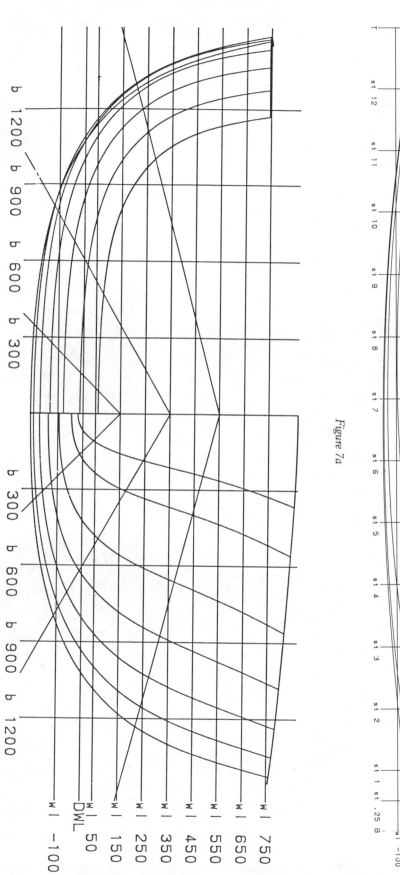

Figure 7b

PHOTO ESSAY

Essentially

I designed Chris Sayer's 6.5 x 3 m Mini Transat single-hander for the two-yearly race from France to the Caribbean. Named *Essentially* after the sponsor, she is a very extreme example of a small racing keelboat.

Chris's Mini Transat, set up ready for planking. Each of the station moulds you see in the photo corresponds to the cross-sections shown on the diagram (Figure 7), less the thickness of the boat's skin.

Essentially in frame. The temporary 'moulds' show how the cross-sections on the drawing form the shape.

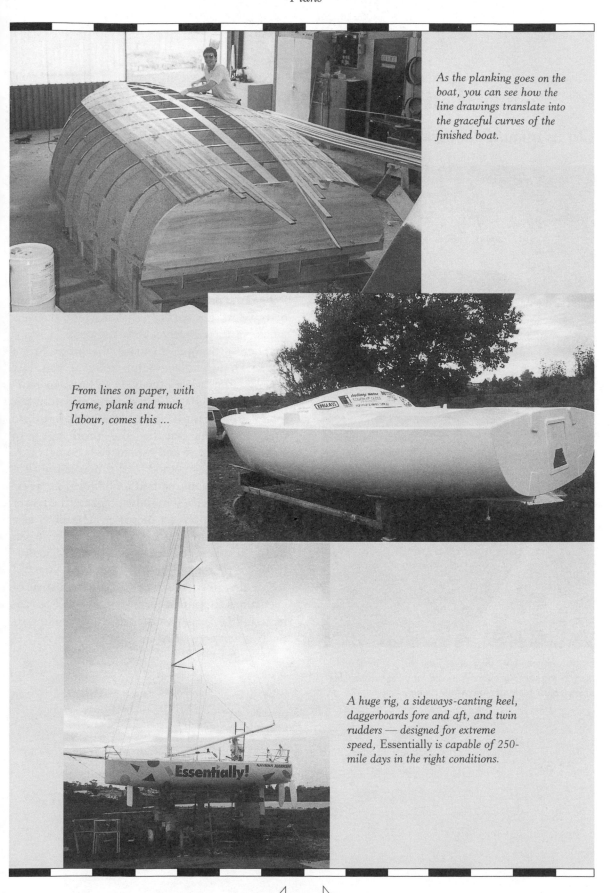

As the planking goes on the boat, you can see how the line drawings translate into the graceful curves of the finished boat.

From lines on paper, with frame, plank and much labour, comes this ...

A huge rig, a sideways-canting keel, daggerboards fore and aft, and twin rudders — designed for extreme speed, Essentially is capable of 250-mile days in the right conditions.

Chapter Seven

Construction Methods

Or, as a multihull builder would say, 'There are several ways of skinning a cat'

Within the range of boat sizes usually attempted by the home boatbuilder, there are many methods or building systems available. While we will not cover all possible permutations in these pages, there are three systems common and particularly suited to the builder possessed of only meagre resources.

The first method is the now traditional system of building a virtually self-sufficient skeleton which is then covered with plywood, a 'skin' that does little more than keep the water out.

The slightly disparaging tone of this last remark is intended; plywood can be an incredibly strong material when used appropriately, but many 'fully framed' boats greatly underutilise these strengths, wasting the effort put into building complete and expensive framing.

However, this method — the most commonly used of the various methods of building plywood hulls — does have some advantages as the size of the considered project increases. The sheer size of the panels involved in a 10 metre boat can make the other methods impractical, particularly for the person building on their own.

I've described the process as a sequence of events in a simplified form so that it will be easier to follow. Note that the real thing would normally have more stringers than I've drawn.

Hobo, *showing how the frames form the interior structure. Hobo uses the outer skin as a fully stressed member, so she's built more like an aircraft than a boat.*

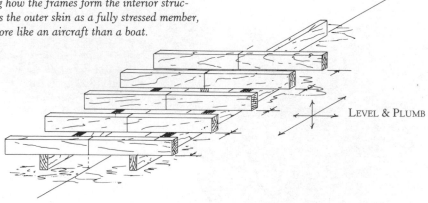

LEVEL & PLUMB

Figure 1: Building frame Type A.

Rogue, *showing the simple jig and the framing set up on the bottom panel.*

1. Frames are built up from the plans, usually from sawn 'solid wood' using plywood gussets at the angles, notched for stringers and keel, and prepared for erection on the building frame. The 'frames' in this instance include the transom, stem, and — in some boats — the centrecase.

2. A building frame, or jig, is built. These vary greatly in type but each serves much the same purpose in that they hold each frame piece accu-rately in place, correctly oriented with regard to the rest of the frame members.

Two commonly used building systems are to set the boat up upside down in either of the fol-lowing ways. Type A (Figure 1) is a floor-mounted jig which uses extended frames to sup-port the framework, over which the stringers and then the planking are fastened. The frames are extended down to the floor or to a level base, and are — when the boat is ready — trimmed off at gunwale height. Type B (Figure 2) uses a

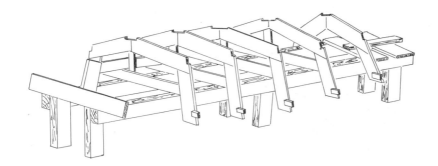

Figure 2: Building frame Type B.

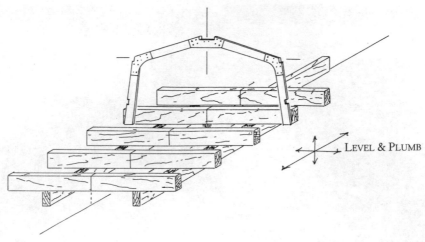

LEVEL & PLUMB

Figure 3: Building frame with midships frame.

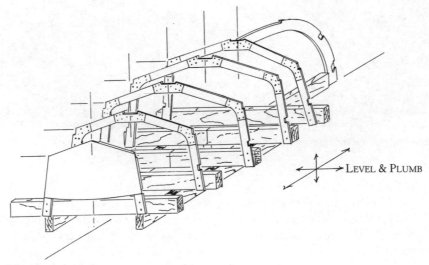

LEVEL & PLUMB

Figure 4: Building frame with frames, transom and stem.

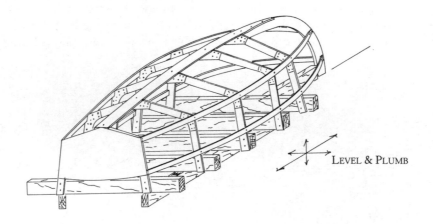

LEVEL & PLUMB

Figure 5: Completed skeleton.

'bench' which provides a level at working height, usually located by a temporary cross-member set at a common height shown on the plans, and a true base upon which the boat's structure can be assembled, without having to stand the frames up on extensions.

Sometimes the building frame (jig) will include quite extensive bracing, with dummy frames and locating points built in. This is particularly the case for smaller boats when several craft are being built in a series from the one jig. This building frame can be complex, and in the case of some boats is the basis for a class 'mea-surement certificate'. This usually is a Type B jig, and (if for a popular design), is often a saleable item.

3. Once the building frame is completed and has been checked carefully for dimensions, level and square the first frame, in my case usually the midships one (Figure 3). (When I get the largest frame erected it gives me quite a sense of progress, helpful in keeping me motivated.) The others are then erected, working fore and aft from midships (Figure 4). This is an exciting time as the shape of your heart's desire gradually

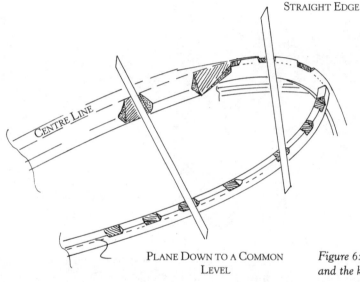

STRAIGHT EDGE

CENTRE LINE

PLANE DOWN TO A COMMON LEVEL

Figure 6: Fairing off (bevelling) the chine and the keel ready for the plywood skin.

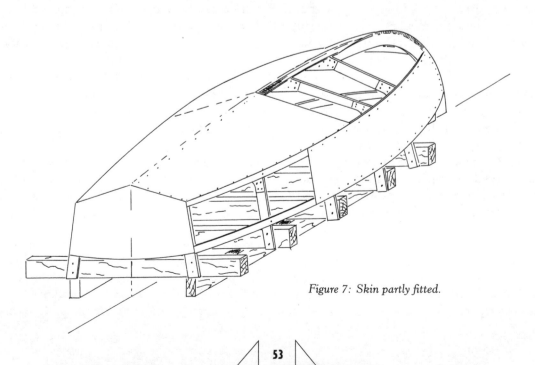

Figure 7: Skin partly fitted.

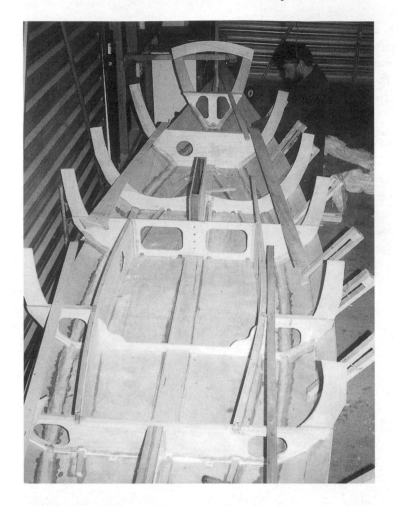

The construction system used for most of my boats. This is Rogue *having the first planks fitted. Note how the seat fronts and centrecase tie the frames together.*

appears before your eyes. This is usually a time when my Dearly Beloved and I sit gazing into the distance, starry-eyed (I do anyway, and she's used to it by now), discussing the wonderful things we're going to do with this new boat that suddenly looks like an achievable reality.

4. With the frames up and the stem and transom in place, it is time to fit the stringers and keel batten (Figure 5). For those who have not had to do it before, their first introduction to bending wood into 'fair curves' can be frustrating. Wood is a material with 'character'. Two long thin bits, both from the same rack at the timberyard and, to the inexperienced eye, almost identical, may behave differently.

5. With the stringers and keel batten in place, the stringers are 'faired off' and bevelled, ready to take the sheet ply 'skin'. This is often done with a power plane; for the less experienced it

pays to use a spokeshave and straight edge to fair a little patch every 300 mm or so, then run along with a hand plane and join the fair patches up. Something I find which helps is to plane a series of short sections with a spokeshave, perhaps 50 mm every 300 mm, then colour the 'correct' angled sections with a felt-tip pen. When you attack the stringer with the plane, the coloured patches greatly aid the builder to estimate the amount of wood left to remove (see Figure 6).

6. The stem needs bevelling off, as does the transom; the angles here are indicated by the stringers, and again you'll need to lay a straight edge on as though it is a piece of sheet ply 'skin' (Figure 6). It pays to close up the indicators as close as 100 mm as you go around the turn of the stem. The bevelling provides the glueing area, known as the 'faying surface', where the ply skin is fastened to the stem and the bottom.

Incidentally, the term 'faying surface' origi-

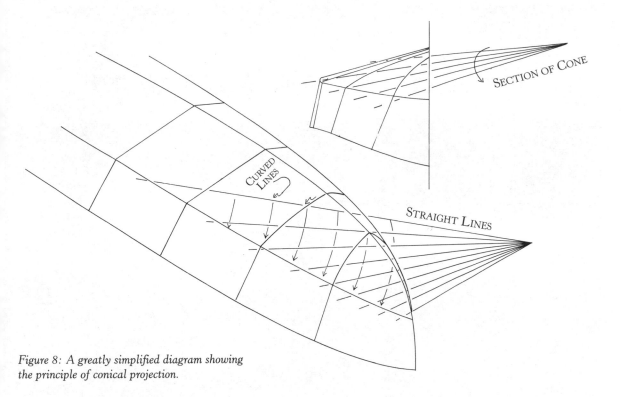

*Figure 8: A greatly simplified diagram showing
the principle of conical projection.*

nally referred to the practice of 'faying': applying a white lead putty to the stem and keel rebate (or 'rabbet') as the solid wood planks were bent in and fastened in place. This is not so common now as the white lead paste is considered too dangerous for common folks to possess, and new solid wood-planked boats are comparatively few. I mention all this so that you can greatly impress friends and loved ones with your depth of knowledge.

7. Having faired off and bevelled the stringers, you'll be facing this wooden skeleton with a stack of plywood at hand and tools at the ready, just itching to fit the boat's skin. Before you charge into it, stop! An understanding of what you are about to attempt will greatly assist you.

Plywood, at least in this type of building, can be assumed to bend in one direction only, never two at once, so you can form simple bends but not compound curves (curved in two directions at once). However, some areas, notably the area around the stem, will appear to be compound curved. Don't be fooled.

When planning your panels, try to fit the tightest curve, usually the bottom panel where

it twists around onto the stem, in as large a piece as you can (Figure 7). The extra leverage afforded by the long panel can make it much easier to achieve what is often a sharp curve. It may pay to cut templates from heavy cardboard or cheap thin ply (try out your supplier for 'coversheets') and test fit them.

This technique, known as 'multiconic' or 'conical' projection, and used by most designers, shapes the boat in such a way that all those graceful sweeping curves are able to be formed from 'simple' curves which are sections of cones or cylinders. If you are going over your skeleton with a long straight edge picturing the series of straight lines across the curved sheet, it makes persuading the sheet ply into your framework much easier (Figure 8).

Note that conical projection is pertinent to almost anything that is developed from sheet materials, whether wood or metal. Metals do have some elasticity though, and designers often use a limited amount of compound curve. The exception is tortured ply construction, a technique closely allied to black magic, which I'll briefly cover in the next chapter.

PHOTO ESSAY

Houdini — from whoa to go

Step one of the new boat — transferring the offsets for the bottom panel onto plywood.

Laminating the stem and building up the transom.

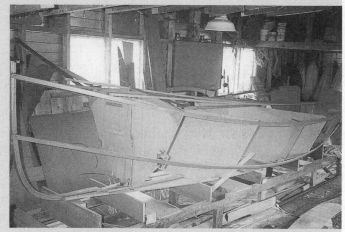

On the jig, the bulkheads and the dummy frames in place, and the stringers on. She's beginning to look like a boat.

The portside topside planking goes on, clamped in place ready for marking.

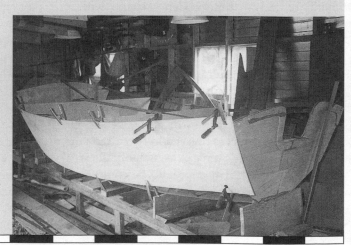

Fitting the 'grounds' for the sternsheets. Note the level. The stern cutout is to fit a short-shaft outboard.

The centrecase is scribed to the bottom. With the stringers in place and the dummy frames removed, the side deck is fitted.

Clamping a panel on to fit the 'deadrise' part of the bottom.

Fitting up the bottom deadrise panel.

All planked up, the centrecase in, the lower chine taped, and the sole support structure partly built.

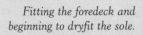

Fitting the foredeck and beginning to dryfit the sole.

Sole and coamings fitted, woodwork essentially complete, and the painting coming along well.

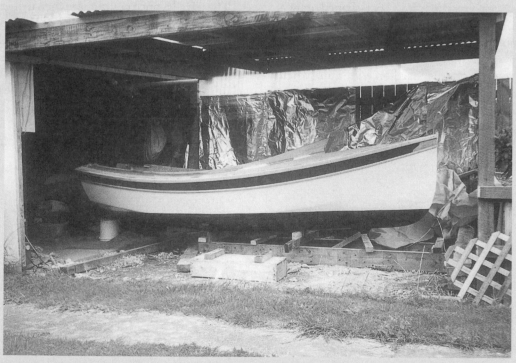

Like a butterfly emerging from its chrysalis, the almost complete Houdini *stands up for inspection. The paint scheme is designed to make her look less high-sided.*

Chapter Eight

Taped Seam Construction

Eliminating a lot of frames and stringers

Taped seam construction, variously known as 'stitch 'n' tape', 'tack and tape', 'flop and glass', 'sewn seam', and other terms on the same theme, is essentially a system of cutting pre-shaped panels, then edge-joining them to form the hull.

Imagine carefully cutting an orange into segments, peeling each one, and pressing each piece of peel out flat. Then imagine taking the pieces and sewing the flattened segments back together. You would once again have a close approximation of the orange's original shape.

In our case the designer has worked out — by hook, crook or computer (I use hook or crook myself) — the shapes of the panels that make up the skin of the boat, and providing you cut the pieces out accurately and join them as per the instructions, you will have a boat-shaped orange!

This is sometimes done with the assistance of a temporary jig to hold the shape until the interior and/or framing is bonded in to stiffen it up.

A variation on this, used on slim hulls where extreme light weight is required, is 'tortured ply', Tornado catamarans and some flat-water kayaks being notable examples. This is an operation fraught with adventure where a great deal

of force is exerted onto thin plywood panels in order to 'torture' them into compound curved shapes.

Earlier, I dwelt on the merits of the common adhesives, and although I have seen phenol formaldehyde (Aerodux or Resorcinol, 'red glue') and elastomeric adhesive sealants such as 3M 5200 or Sikaflex 241 used for this application, epoxy is, to my mind, the best — proven, practical and easily available.

Chapter five told how to splice up the sheets of plywood into long lengths, necessary for all but the smallest of boats, and went on to show an example of a filled and taped join. You've also been shown how tables of offsets and battens can be used to transfer from the plans onto plywood to the full-sized shapes you'll require.

The panels, once cut out, are then drilled with holes in pairs along the matching edges, a few millimetres in from the edges. The panels can then be gradually pulled together by tensioning the 'stitching'; this is achieved with the use of clamps or a Spanish windlass, or any other form of sweat and blood the builder can apply. Once the sewing is complete, frames may be forced into place to help retain the shape. Keeping the

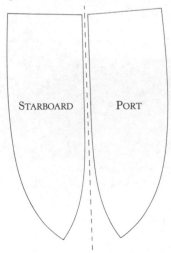

Figure 1: Bottom panels ready for drilling and temporary fastenings.

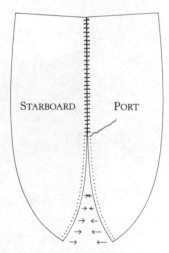

Figure 2: Bottom panels drilled with lacing partly completed.

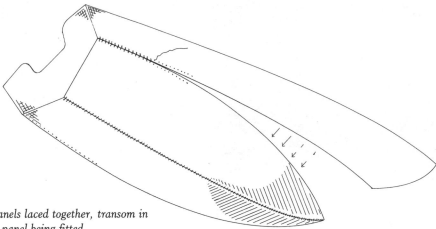

Figure 3: *Bottom panels laced together, transom in place, and port side panel being fitted.*

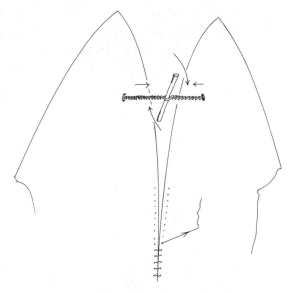

Figure 4: *Using a Spanish windlass to haul the two panels together.*

assembly square and straight can sometimes be a bit of an art! Figures 1, 2 and 3 illustrate the sequence of events as the bottom of a simple hard-chine dinghy is joined.

In case you're wondering, a Spanish windlass is not a dark-haired young woman with a red rose in her teeth who plays castanets in order to summon a breeze for becalmed vessels! Rather, it is a simple loop of rope or cord with a stick through the middle. Put the ends over two objects and haul them together by twisting the loop up with the stick (Figure 4).

Taped seam is mostly used on smaller or light-weight vessels. There are many popular dinghies built using this system, and the strength of the method is such that, properly done, they survive the harsh treatment dealt out to yacht tenders very well.

At the upper end of the scale, quite substantial boats can be built this way, but bear in mind that as the size increases, more and more attention to bracing and holding the shape is required. Above six metres, typically the size of a smaller cruiser, it becomes easier to make the bracing permanent and opt for a thinner plywood skin. Although it is possible to build a boat of much larger size in taped seam construction, the real essence of the system is the unsupported assembly of shaped panels, temporarily held by fastening the edges with the likes of nylon fishing line or soft wire until the join is made more permanent by the application of epoxy resin-soaked fibreglass tape to both sides of the seam.

The resulting shaped shell, often an unbelievably floppy structure, is then stiffened by building in the bulkheads, thwarts, tanks and whatever else makes up the 'furniture'.

Lacing these panels together can be a problem. Applying enough force to the edge of the thin plywood sheet to persuade the said sheet into the highly stressed curve from the bottom to the stem can be difficult to do without tearing the sheet edge. To avoid this it is necessary to spread the loading along the edge, and to ease the curve in gradually. I use monofilament nylon fishing line (get the guy at the fishing tackle shop to show you how to tie knots in it or the job becomes hopeless),

Seagull partly sewn up, with the frames in and ready to 'fill and tape'.

loop of strong string through with pieces of scrap wood on the outsides, and hauling a reluctant curve in that way.

Some taped seam boats need the panel stiffened by glueing a light stringer along the edge before assembly. This is usual in very slim boats, and is intended to give the panel the ability to lie 'fair' when only a very slight curve is being applied. If you are building a short fat dinghy, don't try this unless the designer is positive that it is the way to go. (It probably won't be!)

Joins

It is usual to radius the inside of the corner formed by the two panels with epoxy filler. I use high-density filler, made into a shaped fillet with a rounded spatula, then apply pre-wetted tape while the filler is still wet (Figure 5). Note how the tapes are overlapped to avoid a sudden change in the strength of the panel (known, in engineering parlance, as a 'stress riser'), also

Figure 5: Taped seam, inside.

and 'lace' the seam in 200 mm sections. Each section can be gradually tightened, bringing the edges together a little at a time. This is where the Spanish windlass comes in handy.

Some people advocate the use of soft wire, copper, or a similar easy-to-twist material for the job of lacing. One or two suggest that plastic-covered wire is good in that it will allow the builder to pull the wire out of the glass, leaving the plastic behind. Some use string. The monofilament fishing line I use is simply soft enough to be mown off with the plane when the boat is being prepared for the outside layer of 'glass'. If you've used soft wire it has to be pulled out or it will ruin your plane's carefully honed edge. Take care not to bend the 'tie' around too much in your attempts to remove it; I've seen a frustrated builder break both ends of a wire tie off flush with the surface, and have to chisel it out before the awful mess could be filled and faired off.

As epoxy fills up holes so well I've no qualms about drilling a hole (6 or 8 mm diameter) each side of the area of greatest stress, threading a

Filling the inside of a laced-up chine prior to taping.

Fitting the forward seat tank on Seagull. Note the 'taped' chines that have been filleted with low-density filler prior to sanding. These seams will be almost invisible when finished.

Kindra Douglas's Tender Behind with its taped seams. The next step is to fit the seat tops and paint the interior.

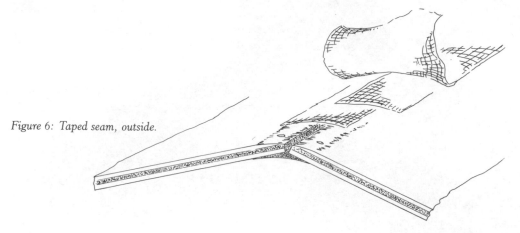

Figure 6: Taped seam, outside.

avoiding an unsightly edge which would be hard to sand off and fill when finishing off your pride and joy.

Once the inside of the corner is sufficiently complete to allow the hull to be turned over, the outside can be given quite a large radius. It is even more important to lay your outside tapes (after bogging up the gaps with filler), in the same overlapping pattern as those inside (Figure 6); the appearance of a line of tape showing through the filler and paint can make an otherwise neat boat look so poor that you wouldn't want to own it.

Sanding off the edges is not easy. The cured fibreglass is usually harder than the surrounding plywood, and there is a tendency to sand a big hole in the wood at the edge of the fibreglass reinforcement. I run along with a fine-set hand plane to take the 'dags' off, and then fill the edge with a low-density epoxy filler mix; sanding this off will be easy, and with care the seam will not show through the paint at all.

Tortured Ply

This method of building caused me to constantly review my store of profanity for the couple of weeks it took to build my daughter's little sea kayak. It was a real fight to achieve the shape I

Daughter Sarina in Snufkin, *her tortured ply kayak, which measures 3.2 m long and weighs in at 5.5 kg dry.*

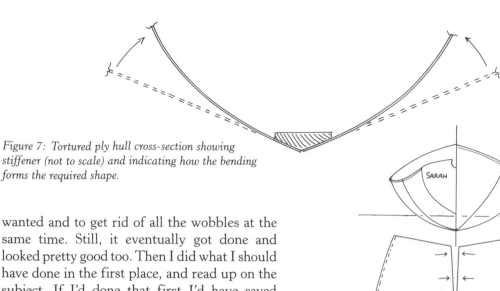

Figure 7: Tortured ply hull cross-section showing stiffener (not to scale) and indicating how the bending forms the required shape.

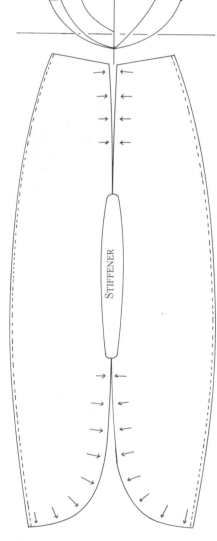

wanted and to get rid of all the wobbles at the same time. Still, it eventually got done and looked pretty good too. Then I did what I should have done in the first place, and read up on the subject. If I'd done that first I'd have saved myself a great deal of hassle. Typical me — if at first you don't succeed, read the instructions! Again, the secret is in spreading the force evenly — only this time there is more force involved and from different directions.

There are two different approaches. One way is to bond the panels together down the centre-line in such a way as to fix the angle of deadrise. This could be accomplished with a wooden stiffener appropriately bevelled (Figure 7), or with multiple tapes and epoxy.

Having created this thing like a pair of squat butterfly wings (Figure 8), the task is then to draw the ends up and the sides in until the hull shape is achieved. The careful use of lacing, gradually retensioning and tensioning again to pull the edges together without over-stressing the necessarily light plywood, does the trick. This method normally has a light stringer glued to what will become the gunwale, thus maintaining the 'fair curve' that you, as a boatbuilder, will be pleased to point out to your uninitiated friends.

The other method is to bond up the matching edges in much the same way, or just loosely stitch them, then slip the lot into a cut-out silhouette of the deck outline and force frames or moulds into the vee-shaped hull to push the sides out into the required approximation of a 'U'. A stringer on the outside gunwale edge can also help a lot with this method.

Generally speaking, the first method forms flatter dish shapes, that is canoes and skinny

Figure 8: Tortured ply light rowing boat, 'in the flat' and the finished shape.

rowing boats, while the second works best with narrow, deep shapes, like multihull hulls.

For all taped seam boats, great attention should be paid to getting the shape of the panels right. The curved edges govern the final shape of your boat, and once the lacing ties are in place there is very little prospect of altering anything without ripping it all to bits again.

Designing your own taped seam boat is a bit of an adventure — determining the shapes and the manner in which they fit together requires patience, some skill, and a lot of tinkering with cardboard scale models. In fact, even if you have purchased a set of plans from a reputable designer, I'd recommend building a scale model using the full-size assembly method. This can

be of real benefit when the full-size boat is under way. Good luck!

One word of warning; the drawings shown in this chapter are for the purposes of illustration only, and their shapes should not be used as a basis of any attempt to build a boat! If you do try, I don't want to know about the results!

In the next chapter, misquoting Frank Sinatra's song, we'll look at doing it 'my way' — my pet method which combines both the framed method and the taped seam system.

Chapter Nine

Doing It My Way

Good ideas I've pinched off other people!

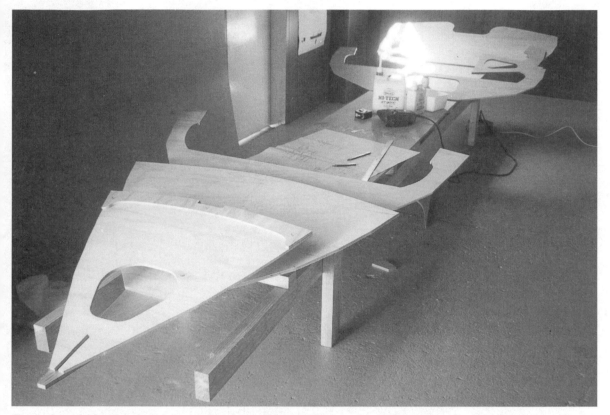

Frames being laid out on the bottom panel as they are cut out. These are for a Rifleman.

Over the years that I've been selling plans, almost entirely to home boatbuilders, I've developed an appreciation of the average handy person's capabilities.

Given encouragement and sufficiently detailed plans, our average handybody (really there ain't any such being — we're all different) is capable of making a good job of even fairly complex projects. However, the thing that stops most people is the length of the job — not in feet or metres, but in months or years.

Time, in today's world, is in short supply. Most of us have to fit our projects in with the demands of family, employment and household duties — and most build because they want a

boat now and can't afford to buy one or pay a boatbuilder. Something that will take years to build is beyond the 'stickability' of our mythical 'average' person. So, going back to my plans — I have taken the best elements from several systems of building, and over the years have developed a building method that, with a modicum of determination and application, will produce a strong, good-looking and functional boat.

My system, not unique to me but in growing use among specialist small-craft designers, is suited to boats ranging from two metres (or less — look at Bob Jenner's tiny version of *Roof Rack*), to perhaps ten metres, and suits a wide variety of types.

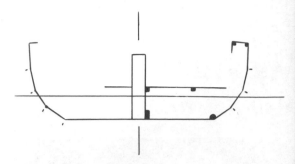

Figure 1: Cross-section of Rogue — *a wide bottom gives the comparatively slim hull enough stability to carry the big sail.*

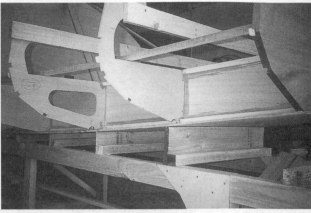

Steve Laker's cabin version of Navigator *with the frames erected and the seat fronts in place. Note how the boat is sitting on its 'jig' of rigid 'ladder rungs' which hold the bottom panel correctly in position.*

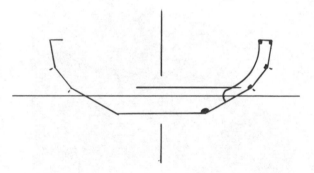

Figure 2: Cross-section of Joansa — *a lot of flare in the topsides accommodates the length of the oars, and a narrow bottom makes her easily driven when used as a sporting rowing boat.*

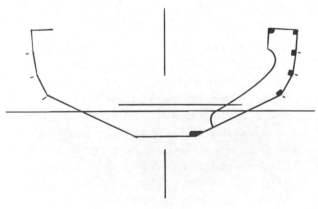

Figure 3: Cross-section of Rifleman — *a narrow bottom panel which forms a planing shoe, steep deadrise to the bilge panels, and a comparatively narrow waterline adapt the shape to a planing light powerboat.*

In order to develop a nice rounded shape from flat sheets of ply, it is necessary to break the curves into a series of straight lines, as in a multi-chine or hard-chine hull. My pet method is to use this 'multi-flats' shape to produce a classic lapstrake appearance.

A builder using this method first makes a series of frames from flat plywood. These have the grounds (basic framing) for seats, buoyancy tanks, cabinetwork and so on, designed-in and built-on prior to assembly. This allows the worst of the often tedious job of building the interior to be done 'in the flat' on a bench, rather than having to build it in place within the confines of the boat. The other major components — transom, stem, centrecase and central spine structure — are also built up, each item a relatively small project in itself and an easily achievable goal, thus helping to keep the budding shipwright motivated. Last on the list is the bottom panel, and in some cases the planks themselves.

With this building method, our small craft — whether a lightweight recreational rowing boat, a runabout, or a sailing dinghy — will use a single plywood panel for its bottom. This panel can be wide, as in a true 'flattie', or very narrow, as in a moderate vee-bottom runabout, and will be much easier to build than the more complex conventional keel. Figures 1, 2 and 3 show how the flat-bottomed multi-chined shape can be adapted to different types of boat.

Steve Laker's Navigator, *all planked up. The bulk of the boat's interior was already in place before the outer skin went on.* Navigator *is usually built as an open boat but Steve wanted a shady place to drink his coffee and read.*

On completion of what is now a pretty comprehensive kitset, the next part of the project is assembling a building jig. As the boat is to be built upright using its bottom panel as a base, all that is required is a rigid 'ladder' with rungs so positioned that the bottom panel is held in place, correctly curved fore and aft, level from side to side, and without twist.

Most of my boats have no structural keel.

However, some have stringers or a keelson attached to the bottom panel for localised reinforcement, and these should be fitted after the bottom panel is properly installed in the jig. If glued to the ply panel when flat, the resulting laminated structure may so strongly resist bending into place that you cannot get enough curve into it to fit the jig!

With the bottom in place and the centreboard

cut out, the 'kit' can be assembled. A clued-up designer will produce a structure that will go together in groups: centrecase with a web frame on the forward and aft ends, stem and spine slotted over the two for'ard frames, transom and after-frames supported by the side seat fronts, and so on. The intention is to have every piece of the interior contributing to the strength of the boat, also helping to hold the frames in place until planking can be fitted — sort of a self-jigging effect!

On the subject of planking, I use stringers to assist the fitting of the boat's outer skin when building very light boats. In *Joansa* I use 20 x 12 mm stringers to keep the 4 mm ply planks fair (free from 'wobbles'). This also applies to boats of a size where a long narrow ply plank may be difficult to handle. On boats with 6 mm or thicker skins, stringers can be done away with as the ply plank, if carefully handled, will lie around the frames smoothly and fairly. This is a process that needs a good eye for a 'fair curve'. This 'eye' can only be developed with practice, so for the novice it pays to spend extra time on this stage of the job.

These boats have to be specifically designed for this building method, and if the designer is expecting a reasonable number of sales they may have built a model or prototype in order to provide offsets for the planking. If this has been done, it greatly reduces the fiddling and skill required to make a good job of the planking — another reason why, on low volume or one-offs, I tend to use stringers. Stringers do assist greatly in planking very light boats, and in many of the smaller boats I use stringers as a matter of course.

Without stringers, many designers, myself included, like to use epoxy filler and glass tape to reinforce the inside of the lapped joint between side planking, and if the butt join (a stitch and taped join in most of my small craft), between the bottom and lowest side plank has no stringer, then multiple tapes inside and out may be required. This is every bit as strong as the other method. I did some trial pieces a while back, and tested them by driving my van over them. I found that in spite of having only one 170 g tape each side of the join and a minimum of filler, the ply always broke outside the taped area; the join itself was strong enough to support my 1.5 ton vehicle!

The bulk of the joints in these boats are epoxy-filled, with the planking joined by butt joints with glass tape or stringer reinforcement, or lapped joints, also with glass tape or stringers.

Holding all this in place until the glue is up to full strength requires a lot of clamps — beyond the resources of many. I use pan head self-tapping screws to hold things together; a piece of ply about 25 mm square with a hole large enough for the screw to slide through goes under the head, the appropriate sized pilot hole is drilled in the area to be held, and bingo — clamped! Note that the outer piece of ply also needs to have a slightly loose-fitting screw hole or you will have great difficulty in screwing the two items together! You can use as many of these as is necessary to hold the job. I also use nylon fishing line to 'stitch' things together; spring, G- and sliding-clamps; scrap timber propped against the jig or workshop wall (don't trip over them, it all flies to bits); nails with pads under their heads to spread the load; bicycle inner tubes with toggles on the ends that act as giant rubber bands; and anything else that might do — let your imagination run riot!

Remember that all of the fastenings, with the exception of the centrecase bolts and a few other highly loaded fastenings in locally stressed areas, are redundant as soon as the glue has gone off. They can be pulled out and carefully put away. The holes are then bogged up with epoxy filler. I've got Scrulox pan head screws that have had a part in holding together a dozen or more boats while the glue set, and will no doubt hold together many more.

If you're going to do a classy varnish job, it pays to beg, borrow or steal enough clamps from friends, neighbours, the brother-in-law, or any other soft touch, as the holes from the PK screws are unsightly if not hidden by paint.

Painting and finishing are, from here on, pretty much the same as normal. Do remember to paint the inside of the buoyancy tanks prior to glueing the tops on though; even using a little make-up mirror doesn't help you to do a thorough job through a 150 mm inspection port!

Should you be tempted to use this method on

Seagull — *the 'shell' is stitched and taped together, the frames are in, and the seats are being fitted (note the footrest being fitted to the end of the sternsheet panel). As you can see, this time I've remembered to paint the undersides of the seats and the insides of the 'tanks'.*

a traditional design, such as one from John Gardner's *Dory Book* or similar, remember that your boat may come out only a third of the original design weight. This, and some other structural considerations that may cause difficulties, leads me to suggest again that you buy plans specifically designed around the method.

Otherwise, the many who have built to this method and have talked to me about it, tell me that of all the boatbuilding systems around, this one particularly suits the part-timer who has a bit of urgency to get out on the water.

Chapter Ten

Making It Look Right

Style is ... (almost) everything

All sweeping curves and rounded edges — Steve Laker's 4.5 m Amy, *a cabin version of* Navigator, *sets off on her maiden sail.*

There are many differences between a 'professional' job and an 'ordinary' one, many little things that separately do not seem important, but together add up to a 'rightness' which is hard to quantify but readily apparent to the eye. Producing this effect is not easy; it requires a combination of engineering and artistry that few other trades require. I will run through a number of tricks that I have developed to add visual class without excessive work.

Most boats are all curves, at home in their environment, graceful to the beholder's eye and fitting with an owner's dreams. Straight lines have little place in a boat; parallel lines and ordinary curves can also look out of place. To produce a boat that 'flows' leads me to sweeping curves, rolled edges and tapers. An example of what can be achieved with even a small boat is the interior of 12 foot long *Janette* (page 114); all curves, no hard corners to produce bruised shins, radiused corners that greatly increase strength, and tapers that reduce the appearance of weight.

This effect is not only two-dimensional. One of the most effective tricks to making a lightweight seat-top look (and feel) better is to double the edge, thus increasing the thickness, and rounding the edge to an elliptical shape (Figure 1). Note that the shape of the rounding over the edge is quite important. It not only looks better than a straight radius but is also much more comfortable to sit on.

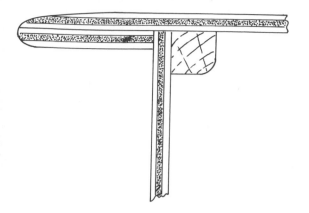

Figure 1: Doubled and rounded-over seat edge.

Tapering the Rubbing Strakes

Tapers are important, often neglected, but very effective in producing an impression of lightness. The rubbing strakes are a prime area for tapering. I often use a pair of outside strakes which contribute greatly to the strength of the boat as well as providing protection from abrasion.

Typically, a 3.5 m boat would use a 45 mm deep by 18 mm thick strake. These look a bit clumsy at the end of the boat though, especially at the bow; tapering greatly enhances their appearance. However, I found that tapering a

pair of four-metre springy strips, particularly getting the reduction in size perfect along the length, is not easy. Like a lot of things, the answer was so simple I couldn't imagine why I'd not thought of it before.

Take the pair of pieces, looking for the point at which the strake will be the fattest — usually the point of the lowest sheer. Laying one on top of the other (on their flat), drive a thin nail through the centre of both at this point. Pull one end across in a scissor action until the reduction in size is right. Pull the other to the right point in the same direction as the first end, clamp at each end, and draw a pencil line along the overlap. Turn over and mark the other, plane to the line, and presto! You have a pair of perfectly tapered rubbing strakes. Make sure you select from stock that will bend easily. This applies to anything that has to 'spring' into an even curve; having got this far in your building project you will be aware of the consequences of a stringer that won't bend easily into an even curve.

It is possible to mount this whole assembly in a vice and plane to shape, but once the pencil line is established I have no trouble in doing each one separately. Note, though, that too much taper is almost as odd to look at as none; in little *Janette* I taper the 45 mm deep rubbing

A new Rogue, *the varnished gunwale and rubbing strake combining with the contrasting colour on the top plank to accentuate the long low, lines of the boat.*

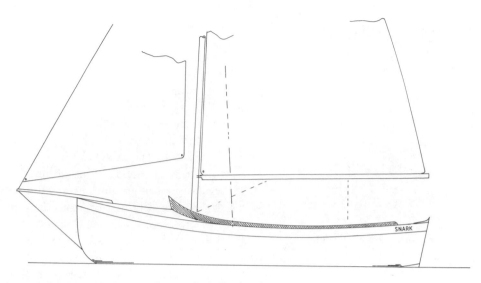

Figure 2: Coaming height tapered to complement the hull's sheerline.

Brian Owen's Maggie Mae *with Wayne Chittenden 'at the helm'. This is an example of a particularly well-proportioned boat. A Peter Culler design.*

strake to 41 mm at the stern and 39 mm at the bow. When fitted with the curved edge downward and sprung into place, it visually 'lifts' the sheer of the boat in harmony with the lines of the clinker plywood sides.

Coamings

Many people can't spell the word, let alone get the proportions right! Another place where straight or parallel lines look slightly 'out', coamings should be long, shallow curves that echo those of the sheer. Even the coaming that runs from the after end of a cabin to the stern, although getting lower all the way, should have a subtle curve to its top and, if possible, its outside edge, that gently mimics the sheerline (Figure 2).

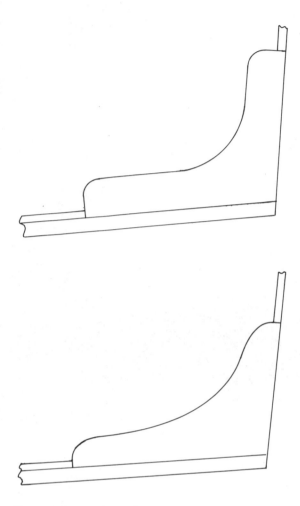

Figure 3: Quarter knees showing poor proportions (above) and good proportions (below).

On an open boat it is possible to make a single sweeping line from the high point of the splash-guard on the foredeck, gently easing down to the low point of the sheerline then sweeping up a little more quickly to the stern's high point — another place for gentle curves which accentuate the lines of the hull.

The smaller the boat the more critical these proportions are. In the small craft which form the bulk of my design work, I have had the opportunity to view several cases of a well-proportioned boat built close to plan alongside one which was built with less regard to the proportions of the secondary structure (interior, gunwales, coamings, and so on). Even with beautifully completed joinery that had been lovingly varnished, the latter still didn't look quite right.

The quarter knees at the transom and the breast hook at the stem are places that need a deft touch (Figure 3). There are not really any rules as to proportions; however, these are not places for semicircles and parallel arms, but for sweeping curves and gentle tapers — stronger as well as lighter, and more graceful.

Colour Schemes

Colour schemes have a major contribution to make to the visual shape of the boat. You can simply paint the whole thing white and have done with it. With a little thought though, a slightly tubby little cruiser, perhaps a bit high in the topsides or cabin for her length, can be visually stretched and lowered; or a rather straight-sided vessel can be changed from having the appearance of a mobile wall into a thing of grace.

Horizontal lines in the topside and on the cabin sides, but following much the same rules for proportions as the coamings and rubbing strakes mentioned above, can draw the eye along the length, rather than allowing the beholder to ponder on the height of the hull, wondering how it can sail without falling over. Such lines can also change the apparent shape of a craft that would benefit greatly from a little extra height in the sheer at bow and stern.

Figures 4a, b and c (page 77) show three variations of colour scheme for the same boat design, and graphically illustrate how the finished job can complement — or ruin — the overall look. If in doubt, do some scale drawings first; if your artistic sense is lacking, ask your spouse and/or kids to look through your pet boat books then colour in your drawings.

Painting

This is the last of the 'building' jobs but one that is often hurried, rather spoiling what would otherwise be a credible effort. Like everything else in this book, I am giving you my own preferences while being aware that there are many other ways — but mine work for me!

A trailerable boat is subjected to extremes of use; it is exposed to a lot of sun, abrasion, water (both fresh and salt), wet leaves on the deck, and a multitude of other detrimental things.

There are two approaches, the first being how I finished my first 'real boat' some fifteen years

Houdini moored for the night in a tiny 'harbour' on Lake Rotoiti. The contrasts of paint from bottom to topsides, together with the varnished woodwork and dark top strake, make this high-sided boat look more in proportion.

ago. *Oliver Zanzibar Zoot Zoot McPhait* (yes, there is a story attached to the name, but the publisher won't give me the several pages it would take to explain it all, and anyway it's not that kind of book!), was a plywood Sigrid class trailer yacht, and the pride of my life. She was first skinned with 170 g glass cloth and Epiglass 90 Rapid, then the whole Epiglass Reaction Lacquer system was applied with a spraygun, carefully sanded with 600 wet and dry every four coats. There were fifteen finish coats and the final surface was rubbed down with Brasso, then silicone polished. It shone!

I came across her a year or two ago, and she is sadly run down, having had little maintenance in the last few years, but the paint is still like that on a new car. Painting her was a major exercise but I can strongly recommend the product.

However, I now tend to go the other way — brush painting, using carefully thinned house enamel or marine-type enamel paint, with a light rub-down of the undercoat prior to the finish coat. (My pet brands are Taubmans Butex

and Epiglass Marinecoat.) This needs a touch-up from time to time and a repaint every three or four years, but my old hatred of having to sand off has been diffused by using a wonderful invention called a Bosch PEX 125 random orbital sander. You can do without one, but you can do the job a lot quicker and easier if you have one.

The point is, you can go high-tech and use complex and expensive paint systems to achieve a very durable finish, or you can use a relatively user-friendly single pot and paintbrush system, and touch it up from time to time. What's winter for anyway? Some people think it's for skiing and other winter sports but you, dear reader, know it's for sitting in your boat, paintbrush in hand, dreaming about next summer.

A few tips though — remember that a brush-mark in your primer coat will show in a nice glossy finish. It is difficult to sand off the hard enamel but the high-build undercoat primers sand easily. A quick whizz over with a sanding block, or a finishing sander using 220 grit, after

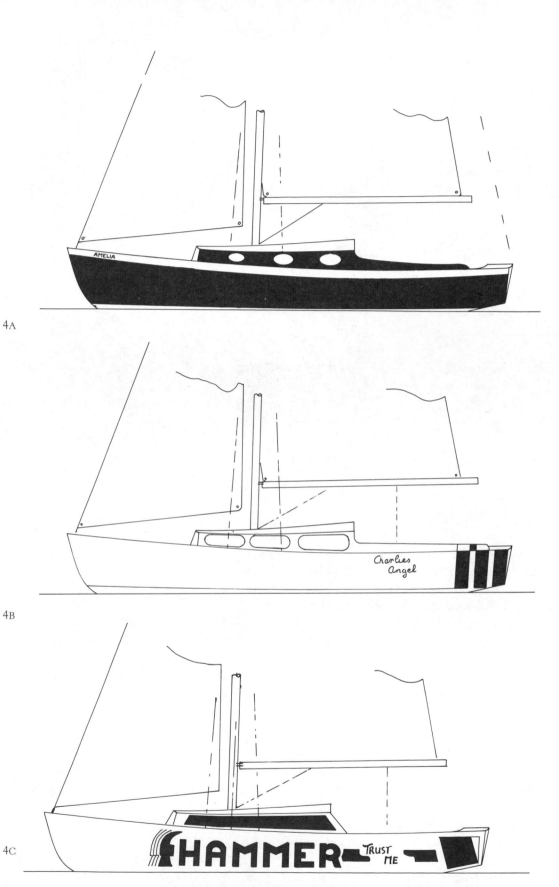

4A

4B

4C

Figure 4: *Graphic illustration of how the finished paint job can complement — or ruin — the overall look.*

Curves are much more pleasing to the eye than straight lines, and this Perth-built Daniel's Boat, *with its carefully worked interior and all varnished finish, is a real credit to its builder.*

your last coat prior to the finish coat, makes a real difference.

Many enamels flow and self-level much better when slightly warm. Try leaving the tin in the sun for a few hours, or putting it in a bowl of warm water before opening. By warm, I mean just touch warm — not hot!

Good brushes are critical. Don't faint when you see the prices; get a couple of the best ones and a big container of thinners with which to clean them. After all, the appearance of several thousand dollars of material, not to mention your labour, is dependent on them. Furthermore, stick to the same brand of paint from first primer to the last touch of the brush.

Varnishing

Lots of people ask me what is the best exterior varnish. My answer? White paint! I only use varnish in small interior or trim areas, the centrecase capping, the tiller, the rudder blade, and so on. It is particularly good on joinerywork which is to be handled a lot, as it copes better

than most painted finishes. However, if anyone really wants to know about varnishing, they should buy Jessica Wittman's excellent book *Brightwork* and leave me with my dislike of the stuff.

Morale

Isn't this an odd subject for a book like this? No fear. Do you dream of a boat on sparkling waters with the sound of the wind in the rigging? Or a boat swinging to an anchor in a mirror-calm cove while the wind howls outside? Or rolling along the old sailing ship routes on your way to a tropical island landfall? Or sitting quietly just off the point with a fishing line over the side? Everyone's dream is different, and buying a set of plans is not just a transaction involving the exchange of paper (money) for more paper (plans), but is the first step on the way to that dream.

Dreams need to be nourished, cherished and cared for, or they die. It is hard to keep the dream in sight over several years, working hard through the day to keep the wolf from the door, while working at nights and weekends to build something you can be proud of.

Any boatbuilding project can be started on about a hundred dollars' worth of plywood and glue plus a little scrap timber — the longest journey starts with but a single step. As the pay-days come and go, a little pot of glue here and another sheet of plywood there can keep you going. As you have more to show for your efforts it becomes easier to budget for the balance. Remember, the biggest investments, and the most valuable, are your own time and skills. And another of my pearls of wisdom for you: the amount of enjoyment you will gain from any given craft is in an inverse ratio to the amount of money you have invested in her.

Building the right boat can be a very reward-ing experience. Alternatively it can lead to bro-ken relationships, financial ruin, mental break-down, and kids who ask their mum, 'Who's that man in the garage?'

Do it right. If you enjoy the work, the family will enjoy your company — when they get to see you. Make sure you spend time with them —-share the dream! Goals and targets govern much of our life; if your goal is to launch your dream, and nothing else will lift your heart, rather than feeling frustrated learn to enjoy the building process.

Do set little goals along the way — build a bulkhead or frame this week, lean it on a wall and get your family to come out and admire it. Make the rudder and tiller early on, then you can sit holding them and let your mind roam the harbour for a few minutes, refreshing the dream.

Get in touch with other builders. Many cities have a boatbuilding co-op which looks for bulk-buying deals, but to my mind the main benefit is the mutual support of the members.

Look after your dream. Don't choose a boat that will take an impossible effort to build. Be open to support and assistance. And above all — have fun!

Chapter Eleven

Camping Equipment

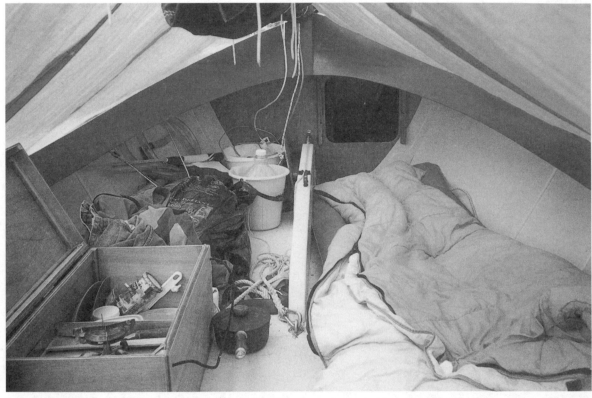

Life aboard Houdini *with the boom tent up. The galley is ready to use, my bed is made up for the night, and my supper is about to be cooked. Very cosy!*

Camping is a matter of comfort. I go cruising to enjoy the sights, smells and sounds of new surroundings, and I find that exploring is much more enjoyable when I am well rested and fed. I also like to be able to set up camp quickly, especially if the weather has caught me out.

One of the great things about camp cruising in small boats is that you have the ability to carry more, and heavier, gear, so this kind of camping can be quite different to others. Comfort is all-important. I can stand being miserable for only a short time, so I have over many years developed my priorities and equipment so that I can travel, if not in five-star luxury, then at least in a style that I can live with.

When planning the equipment, it helps to break the list down into categories — the most important first.

Shelter

If your shelter is inadequate, you may die — a blunt statement, but true. In the open, exposure will kill you faster than anything except a major accident. You can live for a week or so without water, a month or more without food, but only hours if wet and cold. Your shelter should be very quick to erect and be windproof, both in terms of the draughts inside, and not being blown away by the wind.

I have a two-man mountain tent for when I am

alone, and a four-man unit for when there are two. These are the type with fibreglass 'bows' from corner to corner, have two skins (an inner and a 'fly'), a built-in groundsheet, and will stand without pegs once you've put some gear inside to hold them down. I once spent ten hours inside the smaller one, lying in my sleeping bag, with wind gusts flattening the fabric down on top of me — not sailing weather at all!

I do test-assemble my tent on the lawn before I go; a small boat is not always able to keep to a schedule, and one needs to be able to erect the tent in the dark.

Sleeping

A good sleep can make an impossible situation look not nearly so bad. I use an airbed — not your cheapo beach floaty thing, but a 'real' designed-for-the-job one. I carry a small foot pump with which to inflate it; hyperventilation sets in after 100 puffs if I try to inflate the thing, and I have the space, so why not?

Sleeping bags are fine. Bear in mind, though, that it tends to get colder than you would expect when close to water, so take a good one. A cotton liner/sheet will make quite a difference (air it for a few minutes every morning to keep your bed fresh). A blanket or a rug is a worthwhile thing to have, and of course I couldn't sleep without my pillow. This whole lot stows inside doubled plastic rubbish sacks, which in turn are inside a tramper's pack and strapped in under the foredeck. Even after swamping *Houdini*, my bed was dry!

Cooking

Why eat poorly just because you are on holiday? After all, this is supposed to be enjoyable. My principal heat source, after trying many different stoves, is a Primus camping stove of the type that screws onto the top of a '2202' 450 g propane gas cartridge. This provides plentiful heat. I carry a solid 'tablet'-fuelled emergency stove as a back-up, but rarely use it, and occasionally cook on a campfire. Be careful with campfires though — few property owners appreciate the risk.

I now have my stove built into a box, with a stainless steel tray under the burner, and the lid fitted with flaps to form a windbreak. My 'galley

box' is large enough to carry all my cooking gear, including a spare gas cartridge; is solid enough to stand or sit on; and has rope handles, making it easy to carry. Its size is about 700 mm long, 350 mm wide, and 300 mm high. It does duty as a seat, a writing desk, a table, and a washstand. As well as the stove I keep the following items inside:

* A deep-sided 250 mm diameter cast-iron frying pan with lid. This fries, grills, stews and — by putting another container inside with the lid on — bakes. (Makes great bread! Mix the ingredients before you go, then just add the yeast and water when you bake it.)
* A 150 mm cast-iron pot with lid.
* A ½ litre camp kettle.
* A 2-litre billy with lid for big lots of hot water.
* Three small nesting plastic bowls.
* A small stainless steel bowl which does the baking in the big pan.
* Big tin plate.
* Two each stoneware plates and bowls.
* Two mugs.
* Sharp serrated knife.
* Spatula.
* Mixing spoon.
* Wooden stirring and scraping thing.
* Two knives, forks and spoons.
* Salt, pepper, dried herbs — chives, rosemary, bay leaves — brown sugar.
* Matches in a watertight jar.
* Paper towels.
* Dishwashing liquid such as Sunlight — use it to wash your body with too!
* Two dishtowels.
* Scotchbrite scrubbing pad.
* Hard bristle kitchen scrubbing brush.

There's not much you can't cook with the above. You may need to pre-cook some slow things then combine the lot and reheat to serve. A good idea is to set your galley up at home and try out the recipes in the safety of the dining room. With a bit of forethought you'll have great grub when away.

For the menu I use things like UHT custard with those little Christmas puddings you buy in plastic pots. I mix bread and scone mixes and store ready-to-use in Tupperware containers.

Tinned meats, such as Plumrose hams and smoked tongue, are great. Bacon will last a surprisingly long time stripped of its plastic wrapper and kept cool.

I take lots of fresh fruit, UHT milk in small containers (one per day for me), and fruit juice to keep the vitamin C going. A friend of mine, who should have known better, complained to me that her whole family had come down with boils after three weeks away on their big yacht. Scurvy, caused by a lack of fresh fruit, was my diagnosis, and sure enough, the boils disappeared soon after they came ashore and went back to their usual diet.

Eating well is essential. Simple meals are fine, but good quality and a balanced diet are major contributors to a happy cruise.

Clothing

It should be possible to rake up enough clothing from the average wardrobe for at least the first few trips. Loose fits are a help; try for light clothes overlain with warmer ones, and water/windproofs over those. I find that if I can stay warm and dry it's no problem to peel a layer or two off. Any of the now-common synthetic fleeces are useful 'warmers', but bear in mind that they are not usually windproof. Denim jeans, on the other hand, are very poor outdoor wear, particularly when moist.

A really good set of wet-weather gear is important, and some of the best gear I've seen comes from the industrial suppliers like NZ Safety.

Remember that a disproportionately large amount of a body's total heat loss is through the head, so if you're going to buy anything get a snug hat from a ski or outdoor shop. The other thing I wouldn't leave behind is a pair of sunglasses; looking into the sun for hours, particularly when it's reflected off the sea, is a sure-fire recipe for a headache.

Entertainment

I read, on average, a paperback a day when cruising (which would be a worry should I ever contemplate a circumnavigation), so raid the second-hand bookshops on a regular basis. Music is important, so I have a 'sports'-model waterproof radio, as well as my flute so that I can make my own melodies. I also carry a notebook in lieu of a formal log, and I am beginning to draw with pencil and ballpoint.

I carry spares for the essentials, flares, tools, a kerosene (hurricane) lamp, a hatchet, a lifejacket (one for each person on board), a first-aid kit, and a big Swiss Army pocket knife, among other things. Be sensible: you are often out of range of help and supplies, so you need to be pretty self-sufficient — but don't take the kitchen sink (use a plastic bucket instead).

Cruising in small boats is a sociable exercise. I find that in bigger boats I meet very few people; most act as though their neighbours in an anchorage are not there. However, a small boat easing through the crowd to its anchorage, obviously having travelled from afar, attracts people like bees to honey. The number of meals I've enjoyed in return for a telling of my adventures is amazing, and the friends I've made have been treasures.

PART TWO

The Designs

Please note: The plans in this section are not to scale. Please apply to the author for scale drawings (see page 205)

Design One: *Roof Rack*

A practical small yacht tender for a tiny ocean-going keeler

My friend Marcus is something of a free spirit, a bit unorthodox by most people's standards, but thoroughly likeable and very good value.

Marcus had a burning desire to own an ocean-going yacht, something in which he could live and travel to interesting places. In his search for a craft that his meagre budget could afford, he came across a little ferrocement sloop appropriately called *Roc*. This shapely little vessel had completed the Single-Handed TransTasman yacht race a few years before, so had a pretty attractive history.

The tale of how she was refitted and brought from the remote river port where she was lying derelict back to Auckland is too long for this page, but it did involve a complete rollover and dismasting on the bar at Ohiwa on what must have been one of the slowest voyages ever. When Marcus came in and told us of his adventures, we didn't know whether to be wildly entertained or frightened for his safety.

However, at only 5.8 m (19 ft), *Roc* is too small to handle much of a dinghy, but being a deep-keel boat she needs one. Swimming ashore had not proven practical, so I was asked to draw up a suitable dinghy. I'd given it some thought but had been busy on other things when Marcus called in one Friday night, announcing that he was going to build his dinghy this weekend, and was leaving on Monday with it in tow!

I sent him off down to the timberyard for two sheets of 4.5 mm and one of 6 mm ply while I put a fresh sheet of film on the drawing board.

All that weekend Marcus trailed back and forth, picking up the drawings for each new item as he completed the previous one. Monday saw the new dinghy ready for the water. It had only a single coat of undercoat, and no seat top, but as Marcus had only one oar, and that too long, he was going to scull her standing up, so the seat didn't matter for the time being.

It was a couple of months before Marcus turned up on our doorstep again, so I'd been left in the dark about the way the wee boat performed. While relating his latest adventures he told us that the new tender had proven to be a real gem. She towed straight and dry, even in rough weather, rowed well (he'd scored another oar from somewhere), sculled even better, carried three people, and looked a lot cuter than the boxy appearance on the plans suggested.

She hasn't a name as yet, and as I write I'm trying to think of what to call the design. What would you call *Roc*'s 'chick'? I think I'll stick to my original idea of *Roof Rack* — it describes one of the more important aims of the design.

To build *Roof Rack* cut out all of the plywood components, full-sized for the 2.18 m version, or using the shorter lengthwise measurements for the 1.85 m 'shortie'. Make the transoms, leaving the framing around the edges slightly proud so they can be bevelled to take the sides and bottom.

Make the frames as shown on the plans.

Fit the 20 x 20 mm reinforcing to the inside edges of the seat sides; be sure to make port and starboard sides rather than two the same. Fit the seat top making a box section (without the curved ends that brace the transoms), and cut the slots and notches that take the frames and transom framing.

Lay the seat box upside down on a couple of sawhorses, and fit the transoms, taking care to check that they are not twisted. This may take a little juggling but it is what determines the boat's ultimate shape.

Fit the other two frames.

Working from the stern forward, screw the sides to the transom and carefully 'wrap' them around the frames, fastening them to the bow transom. This is difficult if working alone as you must evenly work both around to maintain the shape. It pays to dry fit the sides, check for

Bob Jenner and friends getting his shorter version of Roof Rack *out to set her afloat. Portability was one of the requirements of the design!*

square in both axes, drill all of the screw holes, and then pull the thing apart, apply the glue, and reassemble it.

Fit the gunwales, the end knees on the seat top, then fit the towing eye through the lower frame member of the forward transom. You can flip the assembly up the right way to do this but do wait until the glue is well set.

Paint the inside of the seat assembly; all going well, this will be the last time you ever see this!

Turn the boat upside down again, and — dry fitting again — wrap the bottom panel from one transom over the curve of the bottom to the other end, and pre-drill the screws when you are happy with the way she looks. Refit with glue on all but the joins to the sides.

Stitch the sides to the bottom, fill and tape as shown, screw and glue the twin skegs on, cut out and fit the inspection port into the seat cavity, sand her all off, fit the rowlocks, paint the little critter a nice bright colour, and enjoy yourself. You'll need a pair of 1.5 m oars; she will surprise you with a good feeling of progress when you use them.

Making a Small Splash

Bob Jenner, who'd built a *Rogue* then a *Navigator*, wanted a tiny tender that would fit inside the cockpit of a 5.2 m (17 ft) trailer yacht that he'd bought. There was not much room in the space available, and even the little *Roof Rack* punt would not fit, so after some quick sums we decided to shorten the boat by scaling down the lengthwise measurements to 1.85 m overall.

Bob got the little beast together in quick time, and it was one Friday night that he invited me to another launching party. We met at a pub in Devonport and joined a group of his colleagues and sailing friends who'd also been invited, but who were unaware of the diminutive size of the new craft.

After a drink or two we went out to do the 'job'. Bob's Ford Sierra stationwagon had the new boat tucked in the back, and it was with much laughter from all that we carried her down to the water and shipped the oars for her 'maiden voyage'.

Wielding the two dowelling and plywood sculls mightily, Bob charged out into the swells. It was not long before his concerned frown changed

Three weeks of occasional evenings, about $160, and some bits and pieces left over from other projects — Bob Jenner in his Roof Rack, *enjoying her maiden voyage.*

into a delighted grin, and he surfed up the beach, chattering excitedly about how she went.

Everyone had a paddle, everyone had a good time, and we tried her two-handed out among the moored yachts in a 500 mm chop, making good progress and keeping the intrepid voyagers dry. Back on the beach, all appreciated the joke but were equally appreciative of the qualities of the new boat.

Bob kept the trailer yacht for about 18 months, towing the tiny tender all over the Hauraki Gulf, the dinghy's very light towing weight making little difference to the mothership's performance. He received a lot of humorous comments but all agreed that, although the shape was simple, she did the job she was designed for.

Retirement is a few years away for Bob. Now in his fifties, he goes cruising in a lovely little 7.9 m (26 ft) keeler, and thoroughly enjoys single-handing her up and down the coast. He tells me, with an earnest look in his eye, that he's found the ideal boat, and will never need another. But I've heard that same story before, and figure that a little 8.5 m (27 ft 10 in) cutter that I have on the drawing board might just catch his eye around the time that he exhausts the cruising range of his current boat and approaches retirement. And this time, the full-sized *Roof Rack* will fit up on the deck!

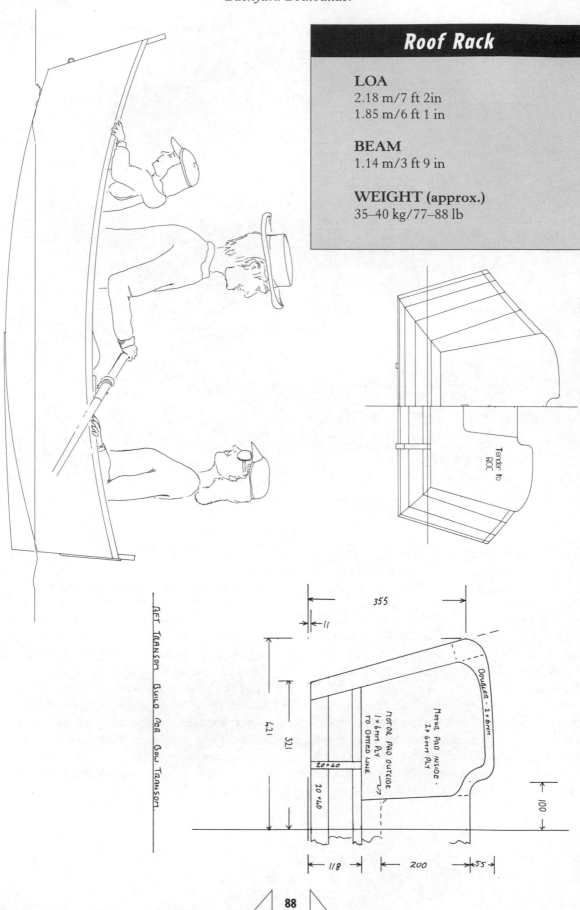

Roof Rack

LOA
2.18 m/7 ft 2in
1.85 m/6 ft 1 in

BEAM
1.14 m/3 ft 9 in

WEIGHT (approx.)
35–40 kg/77–88 lb

Tender to ROC

AFT TRANSOM BUILD PER BOW TRANSOM.

355

11

421

321

20×40

10×40

MOTOR PAD OUTSIDE - 1×6mm PLY TO DOTTED LINE

MOTOR PAD INSIDE - 1×6mm PLY

DOUBLER - 2×6mm

100

118   200   55

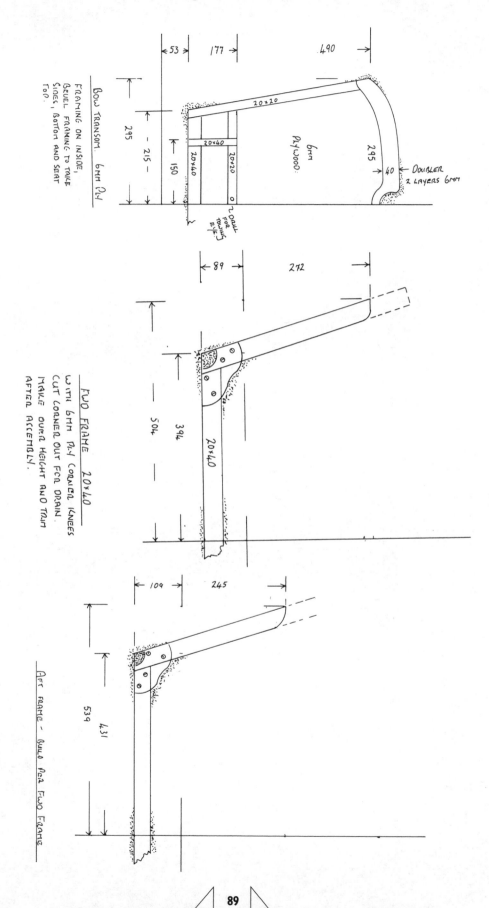

BOW TRANSOM 6MM PLY

FRAMING ON INSIDE,
BEVEL FRAMING TO TAKE
SIDES, BOTTOM AND SEAT
TOO.

490

53   177

295

215   150

20×20

6MM
PLYWOOD

20×40

20×40   20×20

2 DRILL
FOR
TOWING
EYES

295   40 ← DOUBLER
2 LAYERS 6MM

FWD FRAME 20×40

WITH 6MM PLY CORNER KNEES
CUT CORNER OUT FOR DRAIN.
MAKE OVER HEIGHT AND TRIM
AFTER ASSEMBLY.

89   272

504   394

20×40

AFT FRAME — BUILD PER FWD FRAME

109   245

539   431

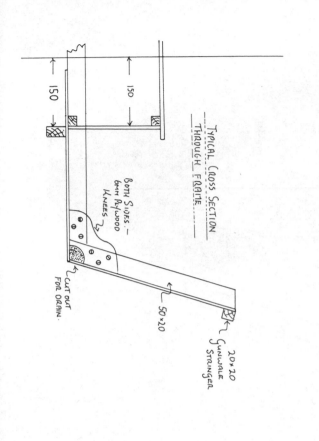

TYPICAL CROSS SECTION
THROUGH FRAME

BOTH SIDES —
6MM PLYWOOD
KNEES

150

150

CUT OUT
FOR DRAIN.

50×20

20×20
GUNWALE
STRINGER

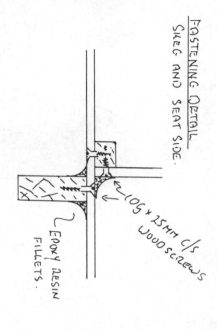

FASTENING DETAIL
SKEG AND SEAT SIDE.

10G × 25MM C/S
WOODSCREWS

EPOXY RESIN
FILLETS.

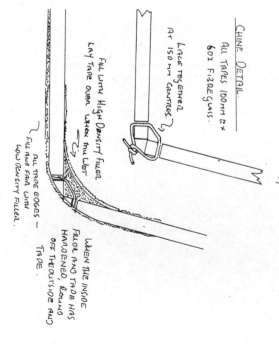

CHINE DETAIL
ALL TAPES 100MM BY
6OZ FIBREGLASS.

LACE TOGETHER
AT 150 MM CENTRES.

FILL WITH HIGH DENSITY FILLER
LAY TAPE OVER WHEN STILL WET

ALL TAPE EDGES —
FILL AND FAIR WITH
LOW DENSITY FILLER.

WHEN THE INSIDE
FILLET AND TAPE HAS
HARDENED, ROUND
OFF THE OUTSIDE AND
TAPE.

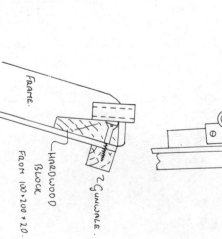

FRAME.

FRAME.

HARDWOOD
BLOCK
FROM 100×200×20.

GUNWALE.

Design One: Roof Rack

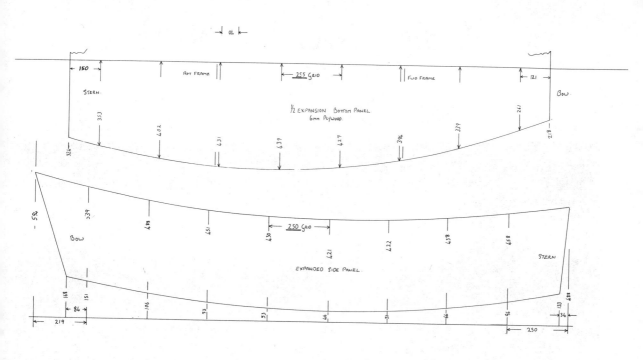

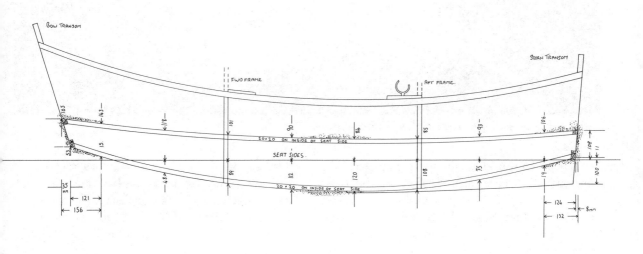

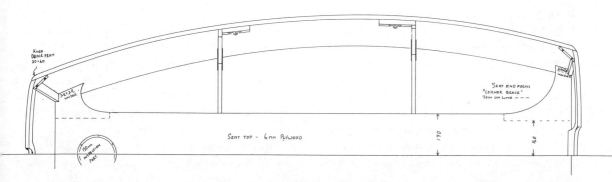

Design Two: *Fish Hook*

A simple skiff

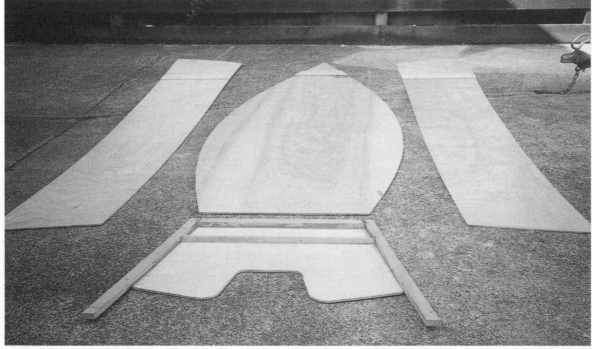

Fish Hook *in the flat, with only the frames and seats to come.*

The idea of a minimum-building-time general-purpose skiff had been lurking in the dim recesses of my mind for quite a while. *Sea Spray* magazine editor Shane Kelly and I had been talking about the best way to set the world to rights, and bewailing the cost and complexity of the plastic or aluminium boats being offered the unsuspecting who just needed something simple from which to dangle a fishing line. My 'idea' surfaced, and we decided that, in order to cure this particular ailment, we would offer an alternative.

I was to produce a design for a simple skiff that could be built in one weekend and finished off during the evenings of the following week. Shane would publish a complete set of plans and instructions over a couple of issues of the magazine, and we could sit back and enjoy a world

that was a little better than when we began.

As so often happens, the best-laid plans of men, mice and editors went awry, the company that owned the magazine decided on a complete revamp, including a change of name, and shortly afterwards Shane was made an offer he could not refuse by a publisher in a completely different field. I was still contributing to the magazine, but it took quite a while for the project to surface again. It all went well in the end. Geoff Green, the new editor, and I got on very well, and the article for the magazine was reactivated; Ancris Associates came up with the glues, resins and surfacing compound; Gordon at Plywood and Marine Supplies donated the ply; and *Fish Hook* was on its way again!

She's not intended to be graceful, but has an air of sure-footed (sure-finned?) safety which her performance lives up to. She can carry a large load in a good chop, and will row better than most things her size. Construction is very easy, and in fact the *Fish Hook* building system was so successful that similar techniques have been included in the sports rowing boat *Seagull*, and the *Setnet/Golden Bay* dinghy.

To build the boat cut out the sides and bottom, joining the panels with butt joins and butt straps as shown in chapter five on basic joining techniques. Make up the frames and transom. Suspend the bottom panel between a pair of sawhorses or boxes, making sure that they are parallel and the bottom has no twist. Dry fasten the frames and transom into place. Then, starting from the after end and working forward, 'sew' the sides into place. As the sides go around, working both at the same time, fastened edge to edge against the bottom and bent around the frames, the bottom will take on the fore and aft curve you need.

As you work forward, it might pay to put a couple of screws first into the transom and then the frames, to hold everything solid as you go. When you get to the bow, check the lot for square and level. At this stage there is nothing that cannot be changed if she is not quite right. Don't be frightened to cut some of the stitching and redo it, as you will have to live with the shape you achieve at this stage.

Back at the sharp end, stitching as before, work up around the stem. Use a plumb bob suspended from the highest point to keep the stem vertical. A stem with a lean on it will ruin the boat's appearance, no matter how well you finish the rest of her.

Your stitching holes should only be about 6 mm from the edges of the panels to be joined; if done this way they are easily covered by the interior epoxy filler and fibreglass tapes.

Put these tapes in place, leaving a gap of perhaps 10 mm between the end of the tape and filler and the frame. When the resin has hardened, take out the frames one by one, apply glue, and replace them, using the original screw holes to relocate them. You can then fill the gaps right up to the frame and tidy up. At this stage you have a boat — a few finishing touches needed perhaps, but she's definitely a boat!

Fit the rubbing strakes or gunwales next; it will help to beef her up for the rough handling that she will receive when you turn her over and back.

Turn her over, round off the chines, and plane off the excess ply at the transom. Tape the outside joins using two layers, and overlap them at the turn of the stem to give extra thickness where she will wear on beaches or concrete ramps. Strength is not a problem with this type of join, but they need to be a little 'overdone' to withstand the abrasion and knocks that small boats sustain.

Fit the skeg(s), screwing through from the inside, using a generous fillet of glue in the angle between the bottom and the skeg. Use good long screws here as the stress of being thumped sideways on the bottom by a wave warrants strong fastenings to keep the boat intact.

The seat tops go in next. If in doubt about your ability to fit the shapes first time, try using cardboard patterns; it is much easier to throw away a piece of somebody else's packaging than to have to buy another sheet of ply. Make little 20 x 20 mm blocks of wood about 100 mm long, and plane them to fit the sides, the top level to take the seat top. These are to support the edges of the seats where they meet the sides.

Remember that the forward and aft seats, if made in two pieces, should be joined fore and aft along the centreline with a 50 x 20 mm block underneath.

Fit the 'corner pieces', breasthook forward and quarter knees aft, then the blocks for the rowlocks, and you are at the tidy-up-and-finish stage. Sand her off, paint her simple and bright, and go boating. You'll only need short oars; she's no speedster — 1.83 m oars (6 footers) will suit for most.

I hope you enjoy this little boat. Building her is about as simple as boatbuilding from scratch can be, and the *Setnet/Golden Bay* dinghy is pretty much the same, as is *Seagull*, later in the book. The few differences are easy to pick up from the plans or from the 'how to' section of the book. If you are building from photocopies and need more detail, it serves you right for being a cheapskate; you would not believe the amount of time and effort it takes to produce a set of plans, and designers — like everyone else — have to eat!

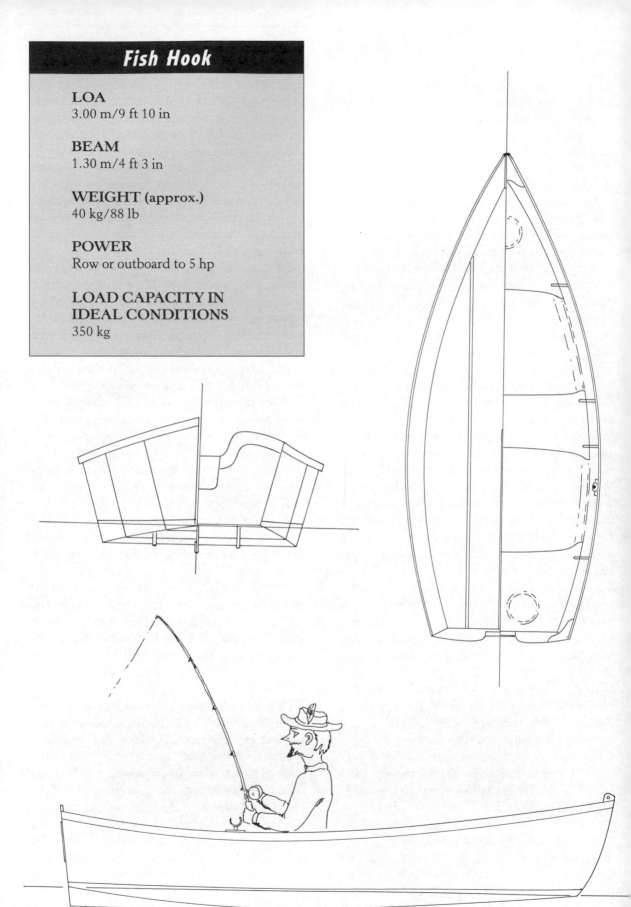

Fish Hook

LOA
3.00 m/9 ft 10 in

BEAM
1.30 m/4 ft 3 in

WEIGHT (approx.)
40 kg/88 lb

POWER
Row or outboard to 5 hp

**LOAD CAPACITY IN
IDEAL CONDITIONS**
350 kg

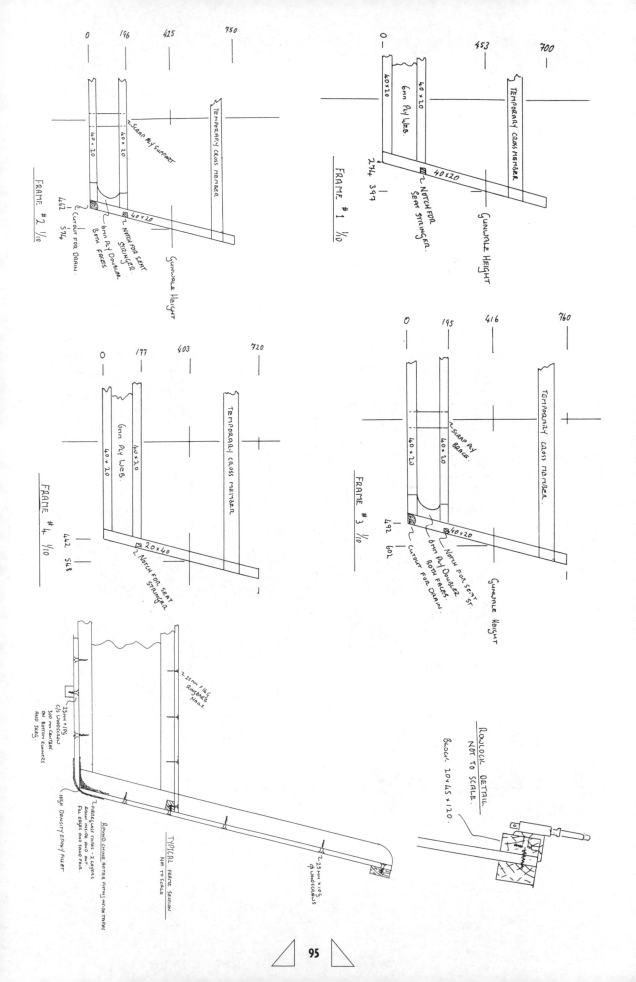

FRAME # 1   1/10

FRAME # 2   1/10

FRAME # 3   1/10

FRAME # 4   1/10

TYPICAL FRAME SECTION
Not to Scale

ROWLOCK DETAIL
NOT TO SCALE
BLOCK. 20×45×120.

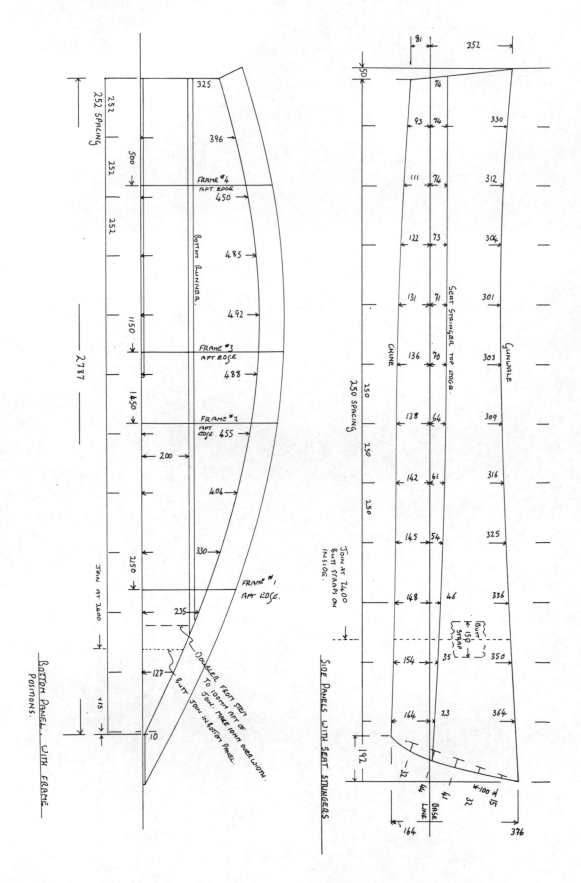

BOTTOM PANEL - WITH FRAME POSITIONS.

252 SPACING

252
252
252

500
1150
1450
2150

2787

325
396
FRAME #4 AFT EDGE
450
485
492
FRAME #3 AFT EDGE
488
FRAME #2 AFT EDGE 455
200
404
330
FRAME #1 AFT EDGE.
235
127
10

BOTTOM RUNNER.

JOIN AT 2400
DOUBLER FROM STEP TO 100MM AFT OF JOIN. MAKE 100MM OVERWIDTH. JOIN IN BOTTOM PANEL.
BUTT
45

SIDE PANELS WITH SEAT STRINGERS

81
352
50
74
330
93 74
111 74 312
122 73 304
131 71 301
136 70 303
138 64 309
142 61 316
145 54 325
148 46 336
154 35 350
164 23 364
32

250 SPACING
150
150
150

SEAT STRINGER TOP EDGE.
CHINE
GUNWALE

JOIN AT 2400 BUTT STRAPS ON INSIDE.

BUTT ⟨150⟩ STRAP

192
44 41 ←100→ 45
BASE LINE
164
376

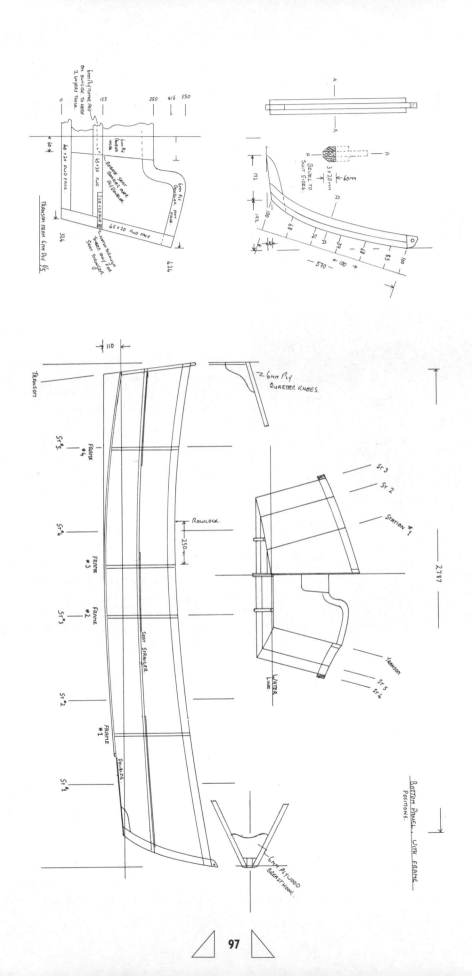

Design Three: *Tender Behind*

A seaworthy yacht tender

Tender Behind's *impressive stability is worth showing off!*

I knew the voice on the phone. Twenty-five years ago, that voice had tried to convince me that A+B=C! And not only that, he'd also informed me that if I were to flick another rubber band in class he'd give me a tender behind!

Now here was Bill Lomas, the maths teacher of my delinquent youth, asking me for a tender behind! After explaining the reason for my merriment and sharing a good laugh, we got down to business.

The requirement was for a tender for his small motorsailer. 'Big enough for Elsie and me, plus some gear and maybe two others at times. Short enough to go on deck, light enough to be easily

lifted up there, and easy for an old amateur to build. Oh yes, we don't have a liferaft on the yacht so it has to be our lifeboat as well!' No longer young, and bothered by severe arthritis in his knees, Bill also needed a boat unusually stable for something so small. Reflecting that life is rarely easy, I told Bill I'd be back to him in a few days.

He liked my preliminary drawings, and we went ahead with what became known to us as *Tender Behind (TB)*, a little boat that has become very popular with many others who have built her. Bill steadily worked away at the new boat, thoroughly enjoying the building and seeing *TB* taking shape. I visited regularly, at first to keep tabs on the new design as it went together, then to visit the man who was rapidly becoming a friend. I found out that he was in fact a qualified naval architect, a fact that he had been loath to mention before as few people appreciate that the design of small craft is not something that a man taught to calculate the internal structure of warships is familiar with.

Bill's invariable good humour and enthusiasm for life made visiting the Lomas home an occasion to look forward to, but sadly he did not live to see *TB* in the water. Bill Lomas died on Christmas Day 1989.

I bought the nearly completed boat from Elsie, and I still have her. She's done many, many miles, carried enormous loads, and done some goodly voyages on her own. I used her at Traditional Small Craft Society meetings one season when I was between 'real' boats, and often took three or four adults exploring several miles up an estuary.

In trials we found that she rowed well for something so short, carried eight people for a few strokes of the oars without swamping, and has stability that has to be experienced to be believed. This is one of the few small boats in

Kindra Douglas's little cruiser. She wanted a boat just big enough for her, a friend, the dog, and a picnic. With the balance lug rig, she is a lively sailer. Note the bamboo mast, yet to be trimmed to the appropriate height.

The Stobbards, father and son, row ashore in Tender Behind.

which one can safely stand up and heave a carton of stores over the rail of the mothership.

Although the rig was intended as one that would never get small children into trouble while sailing around an anchorage, at least two very grown-up 'children' have built their *TB*s ten percent longer (just multiply the fore and aft measurements by that amount; I don't think it would go well if any longer though), and fitted them with a larger balance lug rig. They are using their wee boats for quite adventurous trips; one in fact takes her sleeping bag and pup tent. Cruising comes in many guises, and this might just be the smallest practical cruiser yet.

She sails really well; the wife of one owner tells me that once their 40-footer is secure in its anchorage, the main debate is about who gets to sail the tender and for how long! She tows straight and very dry, motors safely with my 2 hp Honda outboard, and is still light enough to be easily roof-racked on our little car. It's a favourite with the family, and with many others. I've still got mine, and have owned her for longer than any other boat I've ever had.

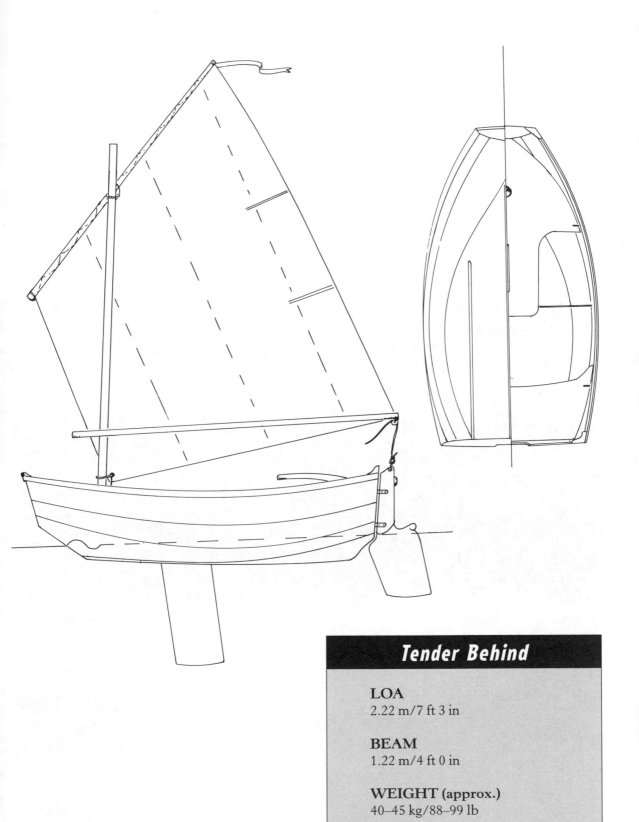

Tender Behind

LOA
2.22 m/7 ft 3 in

BEAM
1.22 m/4 ft 0 in

WEIGHT (approx.)
40–45 kg/88–99 lb

A Day on the Water

The Traditional Small Craft Society had a day 'sail in' planned, a trip up Lucas Creek from Salthouse's Boatyard in Greenhithe to the tidal limit at Albany, some five miles or so away. Although we were without a family-sized boat (I'd sold *Rogue*), we still had *Tender Behind*. However, I had three adults and a child to carry — one too many to be comfortable for the small craft. Frank Bailey offered me one of his plywood kayaks and my wife Jan took that, while our friend Yvonne, daughter Sarina (then aged seven) and I piled into the 2.22 m *TB* with both oars and my 2 hp Honda 4-stroke outboard.

It was a great trip up the creek, with eight or nine boats under sail, oar and paddle. Jan enjoyed the easy paddling of the slim canoe while I rowed our fat little boat out in the channel, making fair progress in spite of the load and using the best of the tailwind and the current to assist us.

This estuary is one of the little-known gems of Auckland. Few houses are visible from the water, with heavy native bush on one side and farmland on the other; there is no traffic noise and lots of birdlife. Winding alongside orchards and bush, scows and steamers serviced Albany township for almost 50 years, bringing out flax, fruit and timber for the markets in Auckland.

Now badly silted up and obstructed in places by fallen trees, there was still enough of the old atmosphere evident to give us a good feeling of the past, and we really enjoyed the gentle drift up with the current and the breeze.

Visiting a township by boat, even one that is a frequent stop by car, is a delight. It feels very different to just climbing out of the car and finding the place there. Tying up at the town's 'back door', chasing the bantam roosters up the path as we went to explore, and sitting on the lawn eating as the children played in the park, was special. Some of the crews joined us on the lawn, some walked over the bridge to the pub for a

Tender Behind, the original, partly built by Bill Lomas and finished off by me.

quick beer, and one or two went window-shopping. We stayed sprawled on the grass watching the sun glittering on the river and the birdlife in the remains of Kell's Orchards.

Back in the boat, a group came exploring with us, up past what was once the steamer wharf, which still shows some of the original wharf timbers at low tide, to the swing basin blasted out below the waterfall. This had been the local swimming hole when I was a lad during the summers of the 1950s and 1960s; there are so many memories here, and even after all these years my name is still visible, carved into the rocks. I remember that the 'big kids' wouldn't let anyone do that until they'd jumped off the top of the falls — high enough to be a bit daunting, even now!

We scraped and bounced over the rocks, going out on the falling tide, past the place where Malcolm Sargent and I had speared an eel so big we couldn't get it up on the shore, and in the battle lost my spear to the monster. I remember my concern was not about losing the spear but about getting out of the knee-deep water before those teeth got me! Talk about having a tiger by the tail — I was terrified!

We were off down the river again, into a headwind this time so it wasn't too long until I gave up on the oars and allowed Mr Honda to keep us up with the larger boats in the fleet. Jan worked hard in the kayak, her first-ever effort other than a paddle along the beach to get the feel of the craft. Downstream we went, past Wharf Road; there's no wharf there now, only memories of fishing with Mum and Dad and my first fish — ever tried to cook a 150 mm sprat? We passed the little inlet where for so many years the derelict ex-Navy MTB (motor torpedo boat) lay, covered with patches of red lead paint where her owners tried to keep ahead of the rot; past the foundations of the old homestead that for years served as a boarding school; and out into Mumbles Reach, where the river widens and the headwind kicks up a real chop.

I took *Tender Behind*, comfortable under power with two adults and a child weighing her down, around ahead of Jan to break the wind and provide a smooth patch for her to paddle in. Friends in other boats gathered around to cheer as Jan used the last of her energy to put the canoe's sharp bow up on the beach, back at our starting point.

We'd had a great day out on the water, four of us in two very little boats, on waters that are on the fringes of the country's largest city — lovely scenery, an enjoyable picnic, lots of memories from my boyhood days, and a first-ever kayak 'voyage' for Jan.

Design Four: *Setnet/Golden Bay* Dinghy

A simple little load carrier, or a sporting sailer with a lot of performance for the money

I was fiddling around with the scale models and drawings for a proposed series of flat-bottomed skiffs when Richard Desborough visited. I've known Dick for more years than either of us care to admit to, and for a lot of those years he has been going to build one of my boats.

Now the fact that the Desboroughs live a good thousand kilometres away means that our contacts are not so frequent, and the ideas as to what he wants have usually changed from one visit to the next. However, on one recent visit I was proudly told that the house renovations were at last finished, and was asked if I could draw a car-toppable sailing flattie of graceful shape that would suit the estuary where the family holidays each summer.

There was a bit of toing and froing as we set the parameters; the size and weight had to suit a compact Japanese car already loaded with four adult-sized people and their gear, and it had to carry the same four on the water in a reasonably comfortable and safe way. Performance was not a criteria, but I've known the man a while, so she has a very high power-to-weight ratio which, allied with the long waterline and slippery shape, gives the boat very exciting performance with one aboard, and more than adequate speed with a heavier load.

Golden Bay is the area in which the 'D's' holiday, hence the name of this craft, and although they did not build her, others have. She sails like a wee rocket. With the large low rig, she carries her sail well (another advantage of this rig is that the spars stow within the length of the boat); this rig is closer winded than many realise, and I know that there are more than a few dinghy sailors around who are embarrassingly familiar with the shape of *Golden Bay*'s transom. I know of one which, in the hands of its eleven-year-old

budding Whitbread star, is a constant headache for the club handicapper, and which, pushed with a 3 hp Mercury, still carries his dad, a mate, two dogs and a pile of gear out to a maimai during the duck season.

Sail her 'free' rather than pinching her; you'll find that you make much less leeway when moving fast. The end result is that you make better progress, and the boat is much easier to control. Note that this type of sail is very powerful off the wind and, rather like many of the old working boats from which she is descended, *Golden Bay* may need reefing for downwind runs when the breeze really gets up.

Setnet is the fisherman's version, without the centreboard, buoyancy tanks and foredeck. She provides a stable and easily propelled fishing platform, one which will stand quite rough conditions in reasonable safety, as long as the user doesn't expect too much speed. She can be built very lightly, enabling someone who doesn't feel like humping heavy weights to slip her into the water; and the same light weight makes her easy to row, and surprisingly fast on an outboard of very small power. A motor of between 2 and 5 hp is plenty for a boat as slippery as this.

Building this boat is just the same as *Fish Hook*. She needs the same care to keep her square and straight as the others in this series of 'jigless' sewn-seam boats, and — being a bit longer — benefits even more from the use of a lighter-weight plywood such as Gaboon or Occume.

I really like these little boats. They are simple enough not to intimidate the first-time builder, don't cost very much, even when nicely finished, and return more fun for the investment than anything else I can think of.

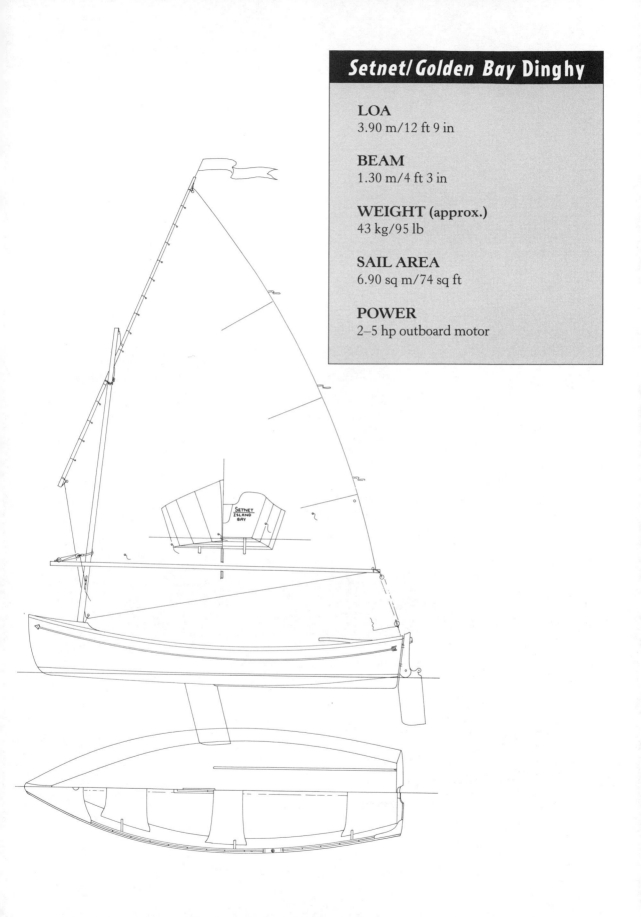

Setnet/Golden Bay Dinghy

LOA
3.90 m/12 ft 9 in

BEAM
1.30 m/4 ft 3 in

WEIGHT (approx.)
43 kg/95 lb

SAIL AREA
6.90 sq m/74 sq ft

POWER
2–5 hp outboard motor

SETNET
ISLAND
BAY

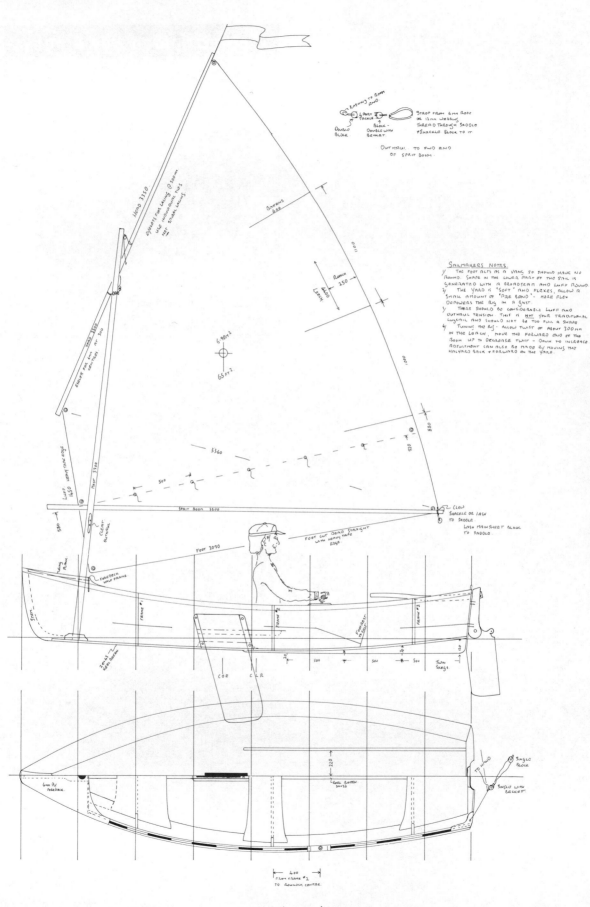

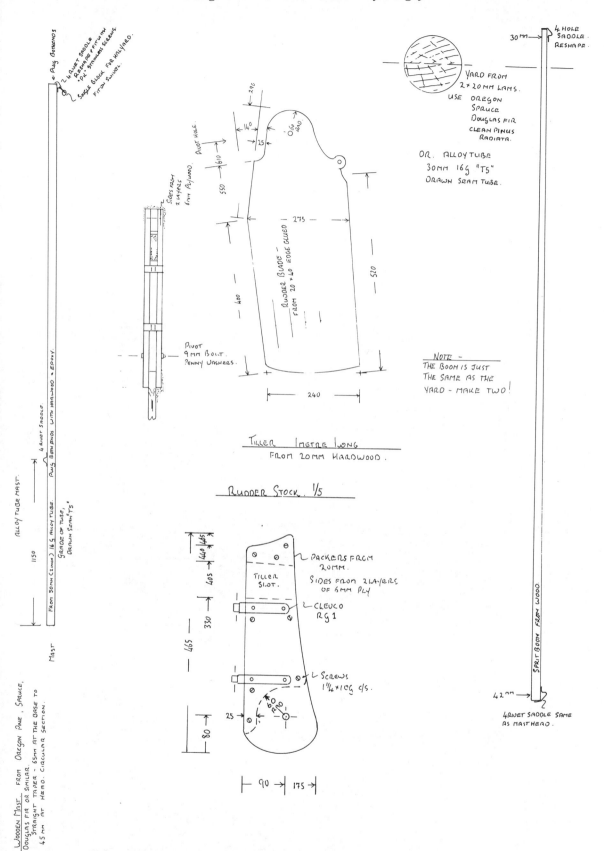

Yard from 2 × 20mm Lams. Use Oregon Spruce Douglas Fir Clean Pinus Radiata.

Or. Alloy Tube 30mm 16g "T5" Drawn Seam Tube.

4 Hole Saddle. Reshape.

30mm

NOTE –
THE BOOM IS JUST THE SAME AS THE YARD - MAKE TWO!

Rudder Blade – From 20 × 40 Edge Glued.

Sides from 2 layers 6mm Plywood.

Pivot Hole.

Pivot 9mm Bolt. Penny Washers.

Tiller 1 metre long from 20mm hardwood.

Rudder Stock. 1/5

Packers from 20mm.

Sides from 2 layers of 6mm Ply.

Tiller Slot.

Cleuco Rg 1.

Screws 1 1/4 × 10g c/s.

Mast

Alloy Tube Mast.

From 30mm (1inch) 16g Alloy Tube.

Grade of Tube, Drawn Seam "T5".

4 Rivet Saddle

Plug both ends with hardwood + Epoxy.

2 - 4 Rivet Saddle. Reshape as fit with Ply stainless screws.

Single Block for Halyard. Fit on swivel.

Plug both ends.

Split Booms from wood.

42mm

4 Rivet Saddle same as masthead.

Wooden Mast from Oregon Pine, Spruce, Douglas Fir or similar. Straight Taper – 65mm at the base to 45mm at head. Circular Section.

1150

200

140

25

610

550

400

275

520

240

465

440

405

330

465

80

25

90

175

R90

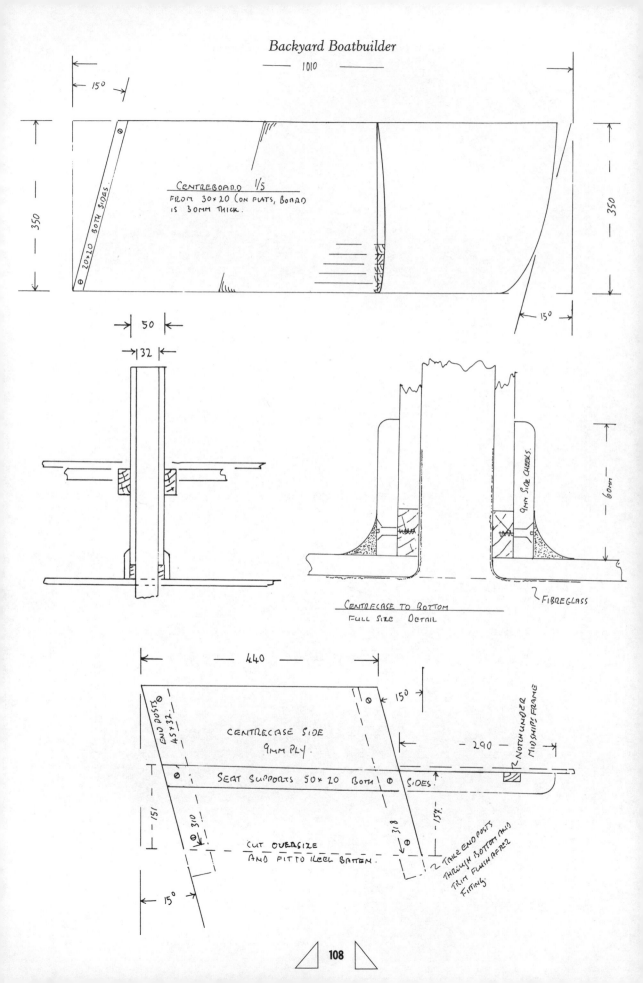

CENTREBOARD 1/5
FROM 30 × 20 CON PLATS, BOARD
IS 30MM THICK.

20 × 20 BOTH SIDES

350

1010

150

150

350

50

32

CENTRECASE TO BOTTOM
FULL SIZE DETAIL

9MM SIDE CHEEKS.

60MM

FIBREGLASS

440

150

CENTRECASE SIDE
9MM PLY.

SEAT SUPPORTS 50 × 20 BOTH SIDES.

END POSTS 45 × 32.

Ø 310

151

318

290

NOTCH UNDER
MIDSHIPS FRAME

152.

CUT OVERSIZE
AND FIT TO KEEL BATTEN.

2 TAKE END POSTS
THROUGH BOTTOM AND
TRIM FLUSH AFTER
FITTING.

150

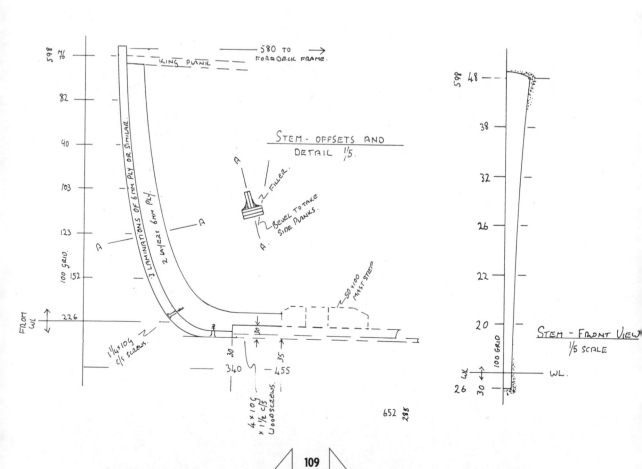

KING PLANK 100 × 20

DOUBLER 9MM PLY

→ 50 ← INWALE STRINGER FINISHES THROUGH THIS SLOT.

80

150 RAD

462 525

338-

20 × 20 AFT FACE.

FOR SETNET - USE ONLY THE PART ABOVE THE DOTTED LINE.

WL

46 136

DRAIN.

FORE DECK FRAME 1/10
6MM PLY
CUT SIDES OVERSIZE BY ABOUT 6MM AND CUT TO FIT. THE SIDES NEED TO BE A LITTLE ROUNDED AND WILL VARY SLIGHTLY FROM BOAT TO BOAT.

598 76

82

40

103

123

100 GRID 152

FROM WL 226

KING PLANK

→ 580 TO FOREDECK FRAME. →

3 LAMINATIONS OF 6MM PLY OR SIMILAR.

2 LAYERS 6MM PLY.

A

A

A

STEM - OFFSETS AND DETAIL 1/5.

FILLER

BEVEL TO TAKE SIDE PLANKS.

A

A

50 × 100 MAST STEP

1¼ × 10g C/S SCREWS.

30

35

340

455

4 × 10g × 1½ C/S WOODSCREWS.

598 48

38

32

26

22

20

100 GRID

STEM - FRONT VIEW
1/5 SCALE

WL

WL.

26 30

652 285

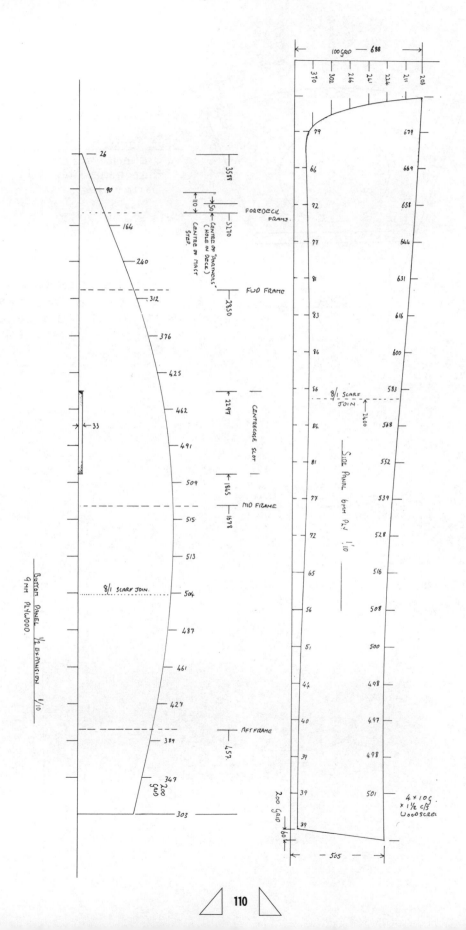

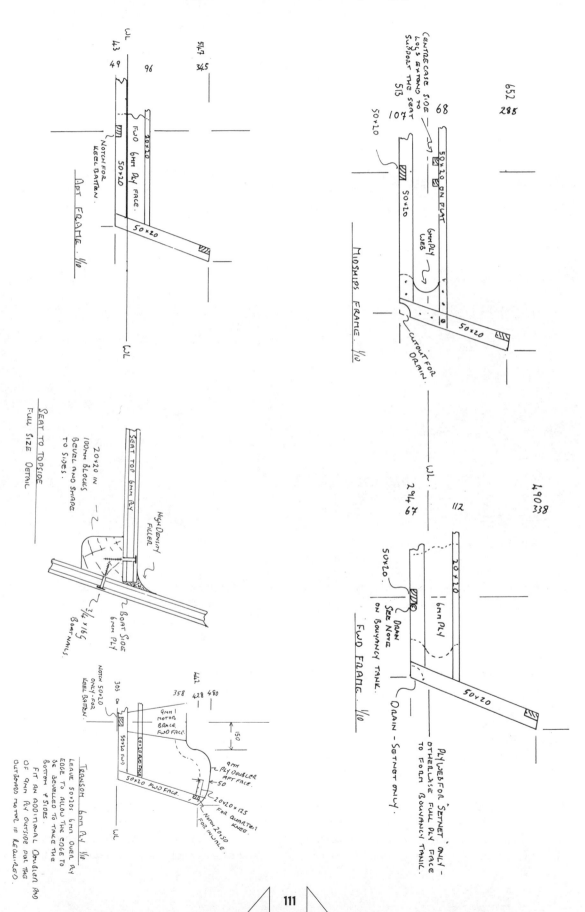

Design Five: *Daniel's Boat*

A happy ten-foot sailing dinghy

The original Daniel's Boat *(and Daniel!) at the Shoal Bay Regatta.*

Twelve-year-old Daniel had pretty much out-grown his Stuart Reid designed miniature flat-bottomed skiff, and his father Wayne Chittenden had arranged to stay with me over one Christmas to do a spot of boatbuilding.

Wayne wanted to build himself one of my light surf dories, and had also asked if I could design a replacement for Daniel's skiff. The brief was straightforward enough — they needed a sailing dinghy light enough to roof-rack, and for a youngster to drag up the beach; but big enough to take three adults sailing, or five when rowing, so she could double as a yacht or launch tender.

She was to be able to cope with the four-mile passage across to the grandparents' place on Waiheke Island; fast enough to be worth sailing in the kids' classes at the annual Waiheke regat-ta; and last, but not least, she had to be built on the thinnest of budgets! By the time my guests arrived, I had a preliminary sketch on paper based on previous boats that were similar but larger.

I gave her more flare than her bigger sisters, to keep the lightweight crew out where the weight would do the most good, lots of freeboard, and a cutout for an outboard motor. The area under the foredeck provides a capacious locker for camping gear, as well as adding to the buoyancy tanks under the seats. She has an underwater shape that allows her to plane when lightly loaded, ample sail area, and reef points for sailing on 'those' days.

Wayne and Dan flew into the building and made rapid progress. I dashed out to the work-

Wayne Chittenden and the completed hull of Daniel's Boat. *Note the locker under the foredeck for buoyancy and dry stowage.*

shop with more drawings as the first parts were completed, and the piles of scrap and sawdust grew as the two boats took shape.

The light dory was completed in time for the Stillwater rowing race. Wayne slid the boat into the water, named her *Emma McLeod* after his grandmother, climbed in, and headed off out to the start line with his (very comely) passenger aboard. He didn't win, but he had a big grin on his face at the finish.

Daniel's Boat went off to their home in Hamilton, structurally complete, but what with one thing and another Dan didn't get her in the water for a while. In the meantime the prototype had created a lot of interest, and over the next few months five sets of plans were sold, the resulting boats being firm favourites with their owners.

That was a few years ago, and then Shane Kelly talked me into the series of 'how to' articles on building small boats that became the basis for this book. I included a photograph of *Daniel's Boat*, and the readers responded with a flood of enquiries, so I redrew the somewhat sketchy plans; here she is.

Wayne and a young friend sailing the original Daniel's Boat at Ngataringa Bay.

Launch day for another Daniel's Boat. This one plies the sheltered waters of Tauranga Harbour.

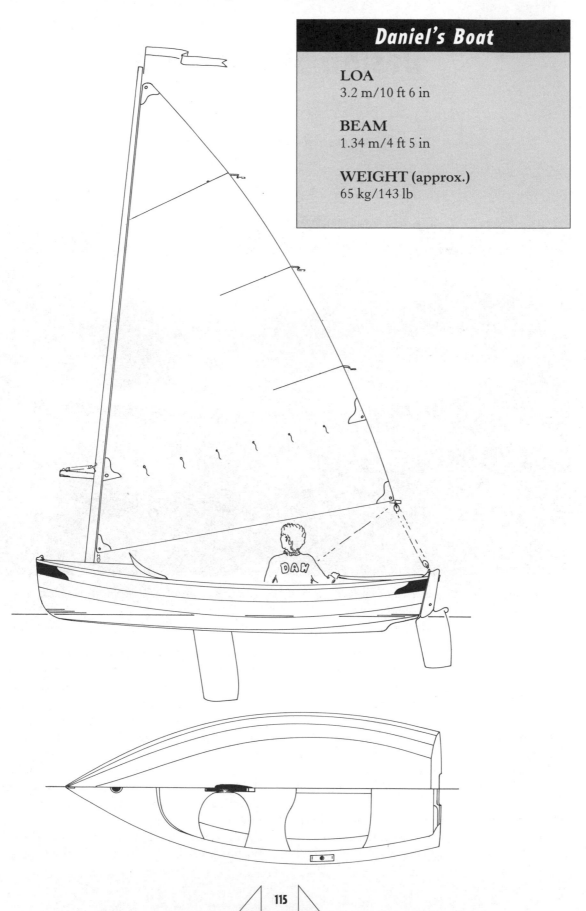

Daniel's Boat

LOA
3.2 m/10 ft 6 in

BEAM
1.34 m/4 ft 5 in

WEIGHT (approx.)
65 kg/143 lb

Design Six: *Janette*

A general-purpose dinghy

A fine example of Janette *by Richard Evans (that's the boat, not the daughter — although she's pretty good too), at Mangere Bridge. Photo: Richard Evans*

The original *Jane*, designed in 1985, sold almost 100 sets of plans, proving the popularity of a simple, inexpensive multi-purpose boat of this size. *Janes* seem to stay with their owners for a long time, suggesting that their virtues outweigh their faults. They are used for all sorts of purposes, from duck hunting to recreational sailing, even by a competitive rower who wanted something more seaworthy than a single shell in which to train on choppy days.

Janette, designed four years and a number of experimental designs later, with the benefit of feedback from *Jane* owners, was intended to provide an improvement on the original, both in terms of performance and building methods.

The aim here was to reorient the boat from a rowing boat that would sail and carry an outboard at a pinch, to a sailing boat that would row, and ditto for the egg-beater. She sails well, enough to get out and enjoy racing in a club dinghy fleet. Surprisingly, she rows even better than the earlier boat, particularly when loaded, and the extra beam makes her a lot better under power.

This boat, like all of the other lapstrake-sided boats featured, is built upright on a simple jig, and is particularly suited to first-time builders. Changes in the construction method make it easy to turn out a really good job without adding to the cost or the weight.

As with many of my customers, John Darling bought a set of plans from me, and with regular contact over some years has become a good friend. His *Janette* came after many years of experience in boating, experience that covered all sorts of craft; the only thing in common was that they were all considerably larger than this one. John rows his some of the time, sails her regularly, and occasionally uses a small motor to take the grandchildren out on the river. He finds that she is smooth and comfortable under sail, but surprisingly quick for her lack of fuss, only outdone by her ability to cover the miles under oars. The boat's performance has been a revelation to him and those of his friends who have not had a chance to try a light and well-proportioned small boat.

For those who want to use a little egg-beater to push them out past the point in pursuit of the elusive fish, for those who find peace in exploring a quiet tidal creek under oars, and for those who want to get away from it all under sail, *Janette* has turned out to be a real friend — simple, inexpensive, and with a real touch of class in her lines.

A Janette *equipped with a second-hand rig. This can often save a lot of money.*

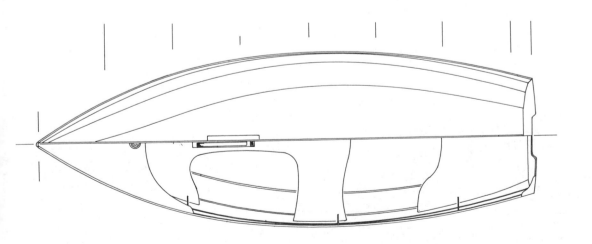

Janette

LOA
3.68 m/12 ft 1 in

BEAM
1.34 m/4 ft 5 in

WEIGHT (stripped)
50 kg/110 lb

SAIL AREA
7.1 sq m/76 sq ft

Design Seven: *Rogue*

The knockabout daysailer with sneaky speed

Steve Humm with his pride and joy, Natalia, *a Rogue design. Photo: Sally Humm*

Many designs start out as an existing idea which is pushed to the fore by a prospective client's enquiry. *Rogue* was one of these.

I'd been experimenting for some time with small cruising yachts, and had built several open boats as experiments in proportion and shape. Some of these were quite specialised; nevertheless, each one had contributed ideas and virtues which I wished to combine in a more general-purpose design.

Bill Harford wanted a big, easily handled outboard dinghy for fishing around the Cavalli Islands on Northland's east coast, while Peter Elstone wanted a lightweight sailing dinghy big enough for two friends with a pile of camping gear.

After the proposal sketches were accepted by the two clients it took longer than I'd have liked to produce the detailed plans, and I was reduced to sending off the plans one sheet at a time. Partway through the process, and right at the beginning of the season, I sold my boat *Hobo*, and I decided to build a *Rogue* in order to have a boat for the summer.

Rogue took eight weekends to build, which included quite a number of midnights (I was in a conventional 45-hours-a-week job at the time). Doing the boat 'in the flesh' enabled me to provide more detail in the plans than was usual, as well as make a complete photo essay of the boat being built. This is not normally possible when working only on the drawing board; it also means that most of the inevitable drafting mistakes get weeded out of the plan set.

Sailing the new boat was everything I had hoped for, and during the year that I kept her she was raced with quite a degree of success, something that has been repeated by several other owners. She acquitted herself well in a long-distance rowing race run on the Weiti River, finishing second in the two-pair oar class (with two passengers — this is not the most serious of events, but once on the water it is hard fought); cruised extensively, at times covering 40 or 50 miles a day along the coast; and got as far afield as Pouto on the Kaipara Harbour, and Cape Colville on the Coromandel.

Not all of this was easy going, and I well remember a long haul into a stiff sou'wester from Pakatoa to my Beach Haven base, some 30 miles from the starting point. It was a long cold day, but I was able to put in for a break and a cuppa at two points, and had an exhilarating thrash through the Motuihe Channel, a place of considerable and well-justified ill repute. By using the reefs as shelter where possible, and sailing free and fast, I was able to dodge my way through the irregular metre-high breaking seas.

In addition to the big adventures, *Rogue* has proven to be ideal for those lovely afternoons

Steve Humm painted the interior of Natalia *a very pale blue to offset glare. We think the splash of varnish looks good too! Photo: Steve Humm*

Steve Humm sailing Natalia *on Lyttelton Harbour. The big rig moves the comparatively slim boat very well, even in light conditions. Photo: Sally Humm*

when being anywhere but on the water would be second-best. As a daysailer she is superb — quickly rigged, fast, and easily handled.

Bill writes that his *Rogue*, powered by a small outboard, has proven to be an excellent fishing craft, coping well with a wide variety of weather conditions, and providing a stable and comfortable platform from which to peacefully wait for a bite.

Another *Rogue* builder, Steve Humm, sails his pride and joy — *Natalia* — on Lyttelton Harbour, and writes that he is delighted with her performance. She's taken a little getting used to, but some good sails up the harbour in the typically stiff easterlies have made him feel at home with the boat's handling.

This design, the winner of the 1988 Traditional Small Craft Society (NZ) beach

cruiser design competition, with its 'classic' styling, comfortable seating, and distinctive traditional rig, has well satisfied those who have built their own *Rogues*, no matter what the intended use.

Among the hundred or more that have been built to date, there are a couple of 'workers' tending marine farms, many family knockabouts, and an increasing number racing in groups around the country.

Rogue also has an increasing popularity in the USA and Australia, which is very satisfying to me as a designer. I, for one, am very pleased with the boat, and would be happy to put the boat forward to refute the oft-repeated statement that boats intended for several different purposes don't do any of them very well.

Rogue

LOA
4.45 m/14 ft 7 in

BEAM
1.36 m/4 ft 5 in

WEIGHT
80 kg/176 lb

SAIL AREA
11 sq m/118.5 sq ft

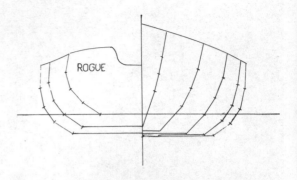

A Weekend Interlude

One Thursday, prior to an Easter weekend, our weather forecasters had been almost gleeful with their predictions of 'high winds, heavy showers, and cool to cold temperatures,' not good for us sailor types, or anyone else for that matter! I admit that Easter pretty much marks the death of our summer, but I'd spent the summer trying to sell goods that people didn't want to folks who couldn't afford them anyway, so badly needed to get away from it all. Even three days would be a welcome break.

The morning of Good Friday dawned just as the Met Office announced it was 'blowing hard from the south', cold enough to convince me that the South Pole was just out of sight over the horizon, with just enough rain to make outside a poor option. My dogs thought otherwise; after a week indoors the two Great Danes left me in no doubt that a long walk was in order. I gave in; two 90 kg dogs pack a powerful argument!

On our way home, at about four o'clock in the afternoon, the dogs and I pleasantly weary from running in the sandhills, I caught a revised forecast that sounded very promising. It was a fast trip home! I dropped the dogs off at their 'babysitters' (greater love hath no man), and checked that another friend was okay to look after the car and trailer while I was away.

Packing a sleeping bag and tent took only a few minutes, as did getting a toothbrush and clothing into a pack. I raided the kitchen cupboards for stores and included my camp stove in the growing pile to be stowed in the boat. With *Rogue* on her trailer and my gear in the boot, we were off to try and catch the tide. We (the boat and I — I like to think we are a team), picked Bill up on the way down to the boat ramp at nearby Little Shoal Bay. At only 80 kg and just under 4.5 m long, *Rogue* took only moments to slip into the water. Her freestanding mast, with its big standing lugsail, was up in no time, and at eight p.m., right on the top of the tide, we sailed out through the lines of moored yachts, gloomy in the orange light from the harbour bridge overhead.

Bill drove the car and trailer back home, and I was soon quietly drifting on the ebb tide through the heart of the city, my way lit only by the glow from office and street lights. It was like being a spectator in someone else's world, an endless anthill agitated by tensions that seemed very remote to me out in the safety of the darkened harbour.

As often happens in this part of the world, the wind had all but disappeared at dusk, and did not fill in again until I was clear of North Head, some two hours later. When it did return it came in over the stern quarter, warm and strong.

With this 'soldiers' wind hustling us along', surfing at well over hull speed much of the time, I decided to head for a favourite bay on Waiheke Island, a sheltered and scenic spot some 25 miles from my point of departure.

At 9.30 p.m. a huge orange moon came up through a wonderfully coloured cloud mass right over the bow, its light making a bright golden path down the passage between the many islands and reefs that lay along my course. This was one of the most beautiful sails I've ever had, the light, responsive boat really moving across a sea of black and gold velvet. There was a real feeling of peace as the bustle of the busy seaport fell behind, to be replaced by silence and the tiny lights of the sparsely settled islands of the inner Gulf.

Awaawaroa Bay, my destination, came abeam as the night breeze died away to a whisper. Creeping silently through the boats anchored in the bay, I felt like a ghost in the sharp black and white of the bright moonlight. *Rogue* moves well in even a very light wind, and we quickly eased our way across the bay to Graveyard Point and my campsite for what was left of the night.

It was about three a.m. when *Rogue*'s stem crunched onto the gravel near a flat patch in the low bank behind the beach. My first action was to set up and start the stove, and while a hot meal was cooking I got the tent up, the bedding out, and the boat anchored off. This didn't take long, and it was with real satisfaction that I reviewed the evening's sail over a fresh cup of tea and a bite to eat.

Next morning dawned fine and flat calm. No problem — this is what cruising is all about! After breakfast it was a snooze in the sun until about ten. Another cuppa followed and then a leisurely breaking of camp, anticipating the arrival of the sea breeze which usually fills in late in the morning at this time of year.

Hauling *Rogue* up to the beach on her long stern line, the gear was soon aboard. The breeze came in just before midday, so up went the big tan sail, and off we went on the next leg of our small adventure. In order to make our planned destination in the strong tides of this area, it was necessary to be in the right place at the right time. Missing a tidal 'gate' could mean another all-night sail to catch up, or abandoning my planned trip right around the island, so even if it meant rowing, it was important to keep on.

The next tide gate was the five mile long passage between Waiheke and the chain of islands to the east. As we made it into the channel the fickle breeze died. With only the tide moving the boat through the air it was just possible to maintain steerage way if I stuck to the current.

I had a great time threading our way through what seemed to be hundreds of yachts, using tiny catspaws of air to slip past boat after boat. Chats with some of the crews were friendly, some envious of my simple freedoms, and some angry that a few scraps of plywood with an old-fashioned rig could pass their expensive status symbols. One thoughtful fellow tossed me some fresh sandwiches and a cold beer as I drifted past his imposing transom, a habit I think should be encouraged!

An occasional few strokes of the oars saw *Rogue* maintain her momentum, ghosting from puff to puff. It was an hour after the turn of the tide when we made the northern entrance to the passage, the beginnings of the flood tide making progress impossible for the moment. Anchored under the cliffs I fired up the little stove for a bite and a hot cup of tea, while watching the fleet of yachts gradually give up and motor in to the clubhouse at the island holiday resort which had hosted the race. About the time I finished my meal a cool breath of air stole across the water, a dark line following not far behind — a private breeze just for us, not meant for the big plastic toys moored outside the by now rowdy beach club, a wind to take me away again to the peace and beauty of the sea.

Fighting the tide and looking for back-eddies and lifts on the current was a challenge, but it wasn't long before we were hard on the wind out on the long swells of the open ocean, powering along in a fresh breeze.

Beating along the steep, rocky cliffs of Waiheke's north-eastern coast, sitting out on the gunwale and crashing through the whitecaps as we tore along, the sunset just beginning to colour the sky ahead, and with my evening destination in sight, I felt as close to Paradise as I could get on this earth.

Garden Cove, my planned stop for the night, is like a giant ice cream scoop out of a line of intimidating cliffs, the narrow entrance hard to find among a maze of reefs. This deep, sandy-bottomed harbour is only the size of a couple of suburban house lots, and is set off by a perfect jewel of a beach that faces directly into the setting sun. It was a doubly welcome sight this evening as the wind died down with the sunset, forcing me to row the last mile or so into the mirror-calm of the little harbour. It had been a long haul!

With the tent up and the boat hauled up on the soft golden sand, I concentrated on the evening meal. They say an army marches on its stomach, and I suspect this is true of mariners as well; it certainly is of this sailor. Dinner this night was fillet steak with mushrooms, new potatoes, peas, and onion rings with cheese sauce. This was followed by plum pudding with custard and cream, the meal helped down by a glass of very good burgundy. Cruising in a small boat doesn't mean missing out on the really worthwhile things in life!

Swimming in the cove that night was great. There was a cool bite to the water, and sheets of bioluminescent light followed my every movement. I was sufficiently inspired to sit on the rocks at the entrance to the cove and play quiet music on my flute as the last of the spectacular sunset faded into a clear night sky. The anthill of Auckland seemed like a distant dream!

Morning — and it seemed that the weathermen were playing jokes on us again! Although the forecast was for a gentle northerly, my observations were that, by afternoon, the wind would probably be from the opposite direction, right on the nose for about 30 miles, and possibly very strong!

Breakfast needed to be substantial, as meals aboard were dependent on calm weather. I demolished bacon, scrambled eggs, cereal, tinned fruit, and a cup of tea, with the rest of the

The view of Rogue *from my tent in the very early morning — quietly waiting for the day's adventures to begin.*

Rogue stopped for a breather in Islington Bay. Note the reef in the sail — testimony to the pretty nasty conditions outside.

pot going into the thermos. While the galley was still unpacked I made some sandwiches for later.

In double-quick time the tent was struck, packed and stowed along with the rest of the camp's gear. *Rogue*, now afloat on the incoming tide, took only a moment to rig, and we were off into what by now had most of the hallmarks of a gathering storm.

In an ideal world I'd have stayed put in my cosy camp until the weather was less inclement, but as one of the worker ants, and one with only a weekend's leave from the anthill at that, I needed to be on my way.

Sailing hard on the wind with a really fresh breeze and no sea running can be fun, and so it was as we tore along the coast. With only a short fetch off the land the wind could not kick up much in the way of waves, and by standing off only enough to minimise the turbulence off the cliffs we made great progress, bringing abeam and leaving behind little settlements with wonderful names — Surfdale, Oneroa and Palm Beach — with the wind gradually increasing as we sailed on.

At about one p.m. we came to the end of Waiheke, the end of my ability to keep the boat upright without putting a reef in, and the end of my willingness to go any further without some more food inside me. Pulling into a sheltered spot and reviewing my position as I ate, it seemed that if we stayed at anchor for an hour or so, then made the short leg across to the notorious Motuihe Channel, we could slip through at slack water low tide, and avoid what was likely to be a very dangerous situation as the tide began to run against the really strong wind.

Rule one for the small boat cruiser: 'If you have to wait, enjoy the break!' Sprawled out on the side seat after a delicious lunch, the world seemed a really good place, even if it was blowing hard enough to knock the tops off the waves and leave streaks of foam in their wake. The sun was shining, and all seemed right with at least my part of terra firma.

Time and tide wait for no one though, and all too soon the sail was up with a deep reef in, the tarp was spread over the camping gear, the bailer was ready to hand, and the anchor weighed. By creeping around the ends of the points, using the sheltered waters in the bays, and close-reaching for power when out in the 'real stuff', it was fairly fast going and not too uncomfortable. *Rogue* was coping very well, but the real test was still to come.

Motuihe Island and the volcanic peaks of Rangitoto are less than half a mile apart. Right across the entrance to Auckland's front door and damming up an enormous tidal pool, they also form a wall across the prevailing south-westerlies. The combination of fast tidal streams, underwater rock shelves, and funnelled winds can create conditions closely resembling the inside of a giant washing machine.

I arrived at the entrance to the passage just before low tide, and hung back in the shelter of a small reef to watch the water and assess the situation. There is another way into Auckland, but it would mean retracing my course and covering another fifteen miles — not to be sneezed at in a boat with a five-knot hull speed! And besides, who can resist a challenge?

As the tide turned the seas eased a little, no longer actively breaking, with the crests a bit further apart but still over one metre high. My hand-held wind meter read 25 knots gusting 35. We could cope with that, so out we went like gladiators into the fray.

I sailed fast, with the sail eased to generate the power needed to punch through the waves, almost close-reaching to windward, the speed

giving control as I dodged around the worst crests and powered through the lulls. Although the boat was not pointing high, our speed meant that we made little leeway, and by working along the shoreline tacking on the lifts and headers from the cliffs on the western side I was able to make good progress.

Part-way through I noted with interest that a group of medium-sized yachts that had followed me into the channel had turned back, decks streaming water, rolling violently as they met the irregular seas of the tide race. I, out of the worst by then, was able to feel pretty smug about how we'd coped with the crisis.

Although we were bounced about a lot, I felt quite safe, and it took only another couple of tacks, timing them to take advantage of the gaps in the waves, to make it around the point into the comparatively smooth water of Islington Bay.

After heaving to for a few minutes for a bite and a breather, I set sail again, putting in a long board across to the smooth water of the weather shore, noting as I went that even now the Motuihe Channel was much rougher, and soon would be an impossible place for a small craft such as *Rogue*.

Sailing into the city along the expensive waterfront suburbs is always fun; just imagine the investment those people have to make in order to be able to watch me enjoying 'their' water. Soon the commercial part of the harbour was in view, with cranes, tugs, and cargo containers with exotic names on their transoms, wanderers of the world resting in my 'backyard' before heading off again to who knows where.

Past the city centre, the wind going down with the sun and the lights starting to glow, reminding me of the voyage out. We were helped under the harbour bridge by the tide, and just beyond made Little Shoal Bay where the boat grounded within a few feet of our departure point.

I reflected on a wonderful trip: two days, two nights, all kinds of weather, about 80 miles, and a visit to a world that is always there but is so rarely seen in our continual scurry to keep up with the bills. It was with real regret that I walked up to the phone booth to ring Bill and ask for the car: 'Bill, I'm back — but I wish I wasn't!'

Design Eight: *Houdini*

My escape machine — the biggest little cruiser

Two views of the nearly finished Houdini. That bold bow is just a metre high, but the boat's extreme beam and sheer are all in proportion, and she can look both cute and capable (like three-year-old Brendan!). Note the foc'sl locker and huge cockpit.

I had planned on using the sports rowing boat *Seagull* as my getaway boat over the Christmas period, but when out with a group of Traditional Small Craft Society members, I was bullied (he offered me money — an unfair tactic), into selling her to a guy who had just rowed the eight miles up the river and back in a little pram dinghy. This left me in mid-November with the prospect of no boat for the summer!

That was Saturday. By Sunday I'd done a stocktake of the materials on hand, checked the space in the workshop, measured my garden trailer, counted how many days until the two important TSCS regattas at the end of January, and designed the boat to suit.

Other considerations were that I wanted to be able to tent her and sleep two on board; she could not be too big for my wife Jan or twelve-year-old

daughter Sarina to sail single-handed, but she still had to accommodate Jan and I, Sarina, and three-year-old Brendan in enough comfort to prevent friction between the junior members of the crew. She also had to be seaworthy enough to cope with my coastal cruising ambitions. As a cruiser she needed plenty of storage space, she needed to be dry and comfortable, and also had to be suited to Brendan's need to scramble around.

Houdini ended up a tad over 4 m long, but a buxom 1.8 m wide. She has a single standing lugsail on bamboo spars (the budget was a major consideration!), and a self-draining floor which runs from the mast step back to the sternsheets (the raised seats aft). The cockpit layout provides a 2.1 m (7 ft) long sleeping space on each side of the centrecase, a big locker under the

foredeck, and more storage under the stern-sheets.

With her wide side decks, ample freeboard, and high coamings, she is a dry boat. Her sharp underwater sections make her stable and much faster than many would expect, while the amount of room is just amazing! One could happily accommodate six or seven adults for an afternoon sail, or four 'big kids' plus a mountain of camping gear for a week away. This is a boat that not only thinks she's a 20-footer, but manages to convince most other people that she's much bigger than she really is.

Houdini has a high power-to-weight ratio which makes her fast, nimble, and exciting to sail, while the wide beam and high freeboard keep her safe if a skipper's enthusiasm overtakes caution.

The building started off well. It took me three weeks of evenings, from when I started that Sunday afternoon, to get the boat planked up, centrecase and anchor well in, and the seats and self-draining floor framed and fitted. She presented no problems in the building, and I was able to correct the plans as I went.

My father died that February, and family con-siderations shelved *Houdini* for a while, then a job came along that I could not turn down but which required a move to Rotorua. All of a sudden there were a lot of jobs with urgent tags on them, and once again *Houdini* had to wait.

The wait was worthwhile though. She was finished in Gordon Newcombe's Plywood and Marine Supplies, the best plywood shop around; Derek Hickman made me an excellent sail, and she was eventually launched into the fresh water of Lake Rotoiti.

There is nothing unorthodox in the structure, and any ordinary handyman ('handyperson' is such a clumsy word — sorry ladies, you are included), could build one of these in short order. I enjoy the building process; the many short stints out in the workshop, away from the stresses of work, are very helpful in relaxing a person who has to be an 'up-and-at-'em' sales-man, no matter how you feel on the day.

In any new boat there is a little of trying some-thing new. *Houdini* has a lot of this, derived from the American Cape Cod catboats to a certain degree. I wanted to avoid their notoriously difficult downwind handling and tendency to be heavy on the helm, so I situated the centreboard

Landfall! Awaawaroa Bay on Waiheke Island — one of my all-time favourite places.

well forward, with a very large skeg and an over-sized rudder aft. This had the added benefit of freeing up much of the space in the cockpit which improved accommodation no end. Oddly enough, the underwater foils, being far apart, seem to contribute to the boat's directional stability, which is phenomenal for such a short, fat little boat.

With the big locker under the foredeck sealed off behind a close-fitting hatch, and a considerable amount of air tankage under the floor and in the sternsheets, she will float very high in the unlikely event of 'canning her out'. But I would warn that it takes a long time to bail such a large-volume boat. I think the dinghy bailer that drains the cockpit floor would eventually empty her, but conditions that would swamp this boat would give even a strong crew a hard time while they tried to bucket her out.

The sprit-boomed standing lugsail rig is a favourite of mine, giving a lot of area for not much money, and short spars that stow within the boat when on the trailer. These rigs are close-winded, easy to handle, have a low centre of effort, and are incredibly powerful when

reaching and running — an entirely appropriate rig for a little boat such as *Houdini*. A gaff sloop would cost twice as much to set up, would have much more windage and weight aloft, and would not be nearly as efficient.

My *Houdini* cost me about $800 to build, plus a few odds and ends of timber and paint — a lot of boat for a small budget, but there were no short cuts, other than perhaps the bamboo which came out of my nextdoor neighbour's hedge!

A new design is often greeted with much interest by my friends, and *Houdini* generated more interest than most. I had lots of visitors who, although they slowed me down a little, were always welcome; the building for me is not a commercial consideration — the social side is important too.

In this case I sold four sets of plans while building, before the prototype was launched. I'd have thought that that might have been a bit risky for something as different from my usual style as this, but she worked out very well in practice, and looks like being the family boat for a long time to come.

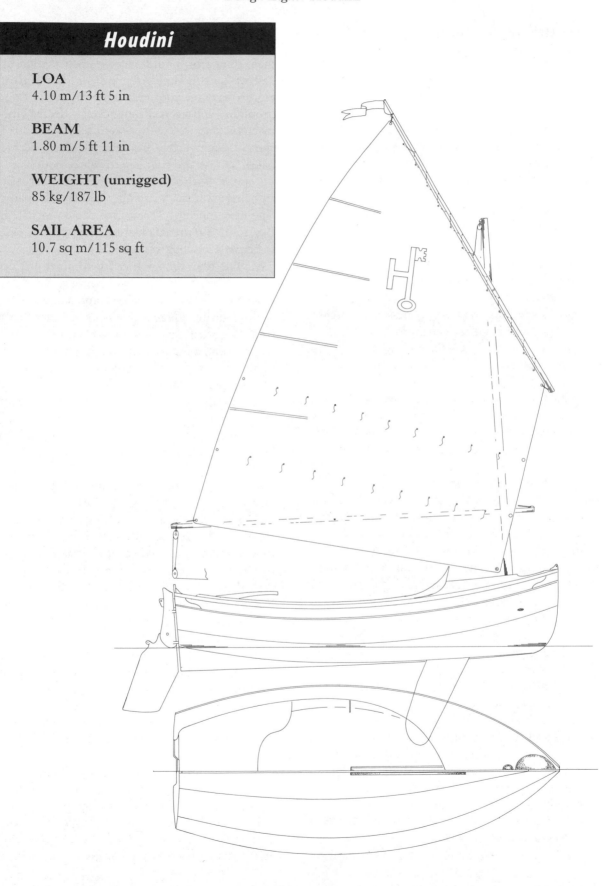

Houdini

LOA
4.10 m/13 ft 5 in

BEAM
1.80 m/5 ft 11 in

WEIGHT (unrigged)
85 kg/187 lb

SAIL AREA
10.7 sq m/115 sq ft

Houdini's Escape
The Perils of Starting in a Hurry

My wife Jan had suggested that I go cruising during the week after Christmas, a suggestion that I seized upon willingly, and the work on *Houdini*, until then sporadic due to family and work commitments, came along rapidly. Even so, it was only two days before we were due to head north out of Rotorua — Jan and the kids to stay with friends in Auckland, and me to cruise the Hauraki Gulf — that I was able to launch the new boat.

In order to at least try her out I took her to the little marina at Lake Rotoiti. It was not a good day as it was blowing a gale, but I was determined to at least get her bottom wet. The simple rig was soon up, the boat organised, and in she went. I gave her a little ceremony:

I name this craft *Houdini*.
May good luck be with her,
And all who sail in her.

She floated very close to her lines, looked lovely in the water, and seemed eager, veering around and tugging on her painter as the gusts tore through the trees around the tiny inlet. I climbed aboard, hauled up the sail, and took off to see how she handled.

In my eagerness to get away I cast off on the wrong tack. There wasn't much room on this side of the jetty, and — Murphy's Law prevailing as always — a gust rounded us up as I headed downwind out of the narrow entrance, putting us under a tree, and taking the mast off a metre or so above the deck! The pride that had me feeling so high as I showed off my pretty little creation came down just as fast as the broken mast, and I was thoroughly chastened as I headed back in.

By the time I got home I'd figured out what to do. A very helpful crew at Mico Wakefield found me a chunk of alloy tube, only a little too short and about the right diameter. The top and bottom of the old spar were cut off and driven into the aluminium section, along with lots of epoxy and self-tapping screws to retain it. Fittings were transferred, a few licks of paint applied, and by the time the evening was over I was ready to pack my gear.

With the celebrations over on Christmas Day afternoon, I slipped away to try and build a boom tent for *Houdini*. Frank Bailey is one of those guys who would have you believe that he has few and modest skills; he is in fact a great guy to have on your side when trying to fix a problem. Between Frank, with his sewing machine, and me, with scissors and a couple of cheap plastic tarps, we created a serviceable boom tent for the boat in a couple of hours.

I stayed with the Baileys that night, greatly enjoying both the hospitality and the company. The next morning Frank came down to Island Bay, in the Waitemata Harbour's upper reaches, to help me launch the boat.

Marcus Raimon, another friend, lives just up from the beach, so there were three of us to get *Houdini* off the trailer, rigged, and launched.

Off cruising for real this time, I headed out from the beach and down the channel towards Kauri Point, a notoriously rough spot where the tidal stream is forced through a narrow place in the channel, and where the cliffs cause the wind to swirl unpredictably. I sailed into this choppy mess just as a squall front arrived. Although I was standing off to keep out of the worst of the land's effects, I was caught 'by the lee' just as a cross sea reared up on the other side and forced *Houdini*'s rail under. She subsided gracefully onto her side, floating high on the built-in buoyancy, but allowing a few bits of gear to float out (gear that I was going to lash in later). Swimming around to the base of the mast, supported by my lifejacket, I undid the halyard and lowered the sail. Back around to the other side, I heaved myself up on the centreboard, brought her upright easily, and a moment later I was aboard, bucketing her out.

It took about ten minutes to bail her out, reef the sail, and get under way again. A crew in a big planing motorboat had kindly picked up my escaped gear, but the weather made it very difficult to transfer it. They passed me a line and gave me a tow into Kendalls Bay and shelter, where we made the transfer.

As the weather was deteriorating further (I'm sure that whoever is in charge of the weather does not listen to the same weather forecasts that I do), I ran *Houdini* up into the bay, taking my tent and wet clothing ashore to wait it out. Although it was blowing hard and the harbour

Waiting for the tide at Kendalls Bay, in the Upper Waitemata Harbour.

was a harsh mix of grey and white, my camp in among the flax bushes was well sheltered for the time being. I washed my salty clothes in the creek, spread them out to dry, walked out the five km to the shops to pick up some more reading matter, and then settled down to await the weather's pleasure.

By late that afternoon, my mountain tent was being completely flattened by the gusts. Only me and my gear inside stopped it from being blown away altogether. *Houdini*, up high and dry as the tide fell, was blown over on her side by the force of the wind. I retrospectively blessed my earlier dunking for forcing me ashore.

It was late the following night, three books and a couple of long walks later, that the weather moderated enough to allow me to carry on down the harbour. I'd been packed and stowed since mid-afternoon, anticipating the wind's lessening as the sun went down. In about fifteen knots of breeze I hauled the boat out along the anchor line, laid at low tide in what would be

deeper water in the flood tide, then set off.

Although I was bucking the incoming tide, I had a quartering tailwind to help me on my way down the harbour and underneath the bridge. The roar of the city, even this late at night on a holiday weekend, was almost overwhelming; it was like a monster in a cage, the sound reeking of stress and frustration. But it's always interesting sailing along the commercial part of the wharf district, the big freighters like giant seagoing horses among the ponies and cobs, the smaller ships that service the coast and harbour. Each class of vessel seems to have a personality; the tugs like short fat hospital matrons, bustling about, nagging and cajoling their charges into their berths; the modern ferries like young women, fashionably dressed and striding purposefully across the harbour; and the giant ocean-going freighters dozing at their wharves while dreaming of exotic ports and faraway places.

It was on the very last of the breeze that I slipped in among the yachts moored at Okahu

Bay in Auckland's centre and put the boom tent up, more by feel than by sight. It was very quiet even though the road was only 50 metres or so away. Making my bed and cooking a meal on board for the first time went about as planned; the order of storage in the big forward locker, and the plywood box galley, with its permanently fitted stove and racks for pots and pans, made the conversion from an open sailing dinghy into a cosy cruiser only a ten-minute affair.

After a comfortable sleep and a good breakfast, it was up sail and off for Waiheke. It's quite a way across to the island from Okahu Bay, and I was aware that I was sailing a boat design that was still untried. *Houdini* is the original, and this was truly a maiden voyage, so it was off along the coast rather than straight out into the channel. The boat felt good in the quartering tail wind as we went across the mouth of the Tamaki River and through the channel between Musick Point and Browns Island. Even with the wake of some very inconsiderate launches tossing us about all seemed well, so the bow was turned downwind towards Waiheke, a dark line on the gleaming sunlit water, and we set off in the wake of the big vehicular ferry heading towards Kennedy Point, some eight miles away.

Although a big jump in a brand new little boat, I felt comfortable; the boat seemed well balanced, stable in spite of the upset a couple of days before, and an hour and forty-five minutes later, with near-perfect conditions, we were inside Putiki Bay, tacking back and forth up the narrow channel to Ostend.

I'd got far enough up the creek to spot the crossed yards of the tiny brigantine *Coriolanis*, parked in her mud berth in the mangroves, and had been wondering where my friends Rhonda and John Griffith would be, when I spotted a bright red beard and floppy hat standing alongside a decidedly feminine form. They'd just remarked, 'Wouldn't it be nice if someone we knew sailed in for a visit,' when I'd come into view around the corner, the jaunty sheer and tan sail of *Houdini* marking her in their eyes as a John Welsford boat. They later told me that they'd decided it was someone else in one of my boats because it had an aluminium rather than a bamboo mast, which meant I had to tell them the apocryphal tale of Friday's launch, the over-

hanging tree, and the fate of the original spar.

After lunch together we went out for a sail around the bay. We found that although the boat was small, there was plenty of room for the three of us, even with all the camping gear aboard.

I needed to be away to catch the tidal 'gate' into the Waiheke Channel. To try and sail through past Orapiu after the tide had turned would take forever, so by two in the afternoon I was out past the point in the deeper water, and off down the coast again.

Scooting down past Rocky Bay, Awaawaroa Bay, and Big Muddy Bay, I was soon into the channel, well in time for the tide, and getting a feel for how quickly the little boat could cover the ground.

Cruising in little boats is deceptive. I've been fooled several times at the speed with which a passage can be made. In fact, on one memorable occasion, I got myself comprehensively lost off the Northland coast after overshooting my landfall by about twelve miles, and ending up on a piece of coastline that didn't look anything like where I wanted to be! I still haven't got a log; a mechanical one is too vulnerable on a boat that is beached a lot, and the lack of an electrical supply makes an electronic unit impractical. But the rate at which the price of hand-held GPS (Global Positioning System) units is falling might provide a solution to several navigational problems in the near future.

Waiheke's Bottom End is one of the most beautiful parts of the Hauraki Gulf, a stunning cruising ground no matter which part you are in, and sailing *Houdini* along the many little bays, ducking in to look over an interesting boat or crossing the channel to check out a possible overnight anchorage, was like being in heaven. As the clock crawled around to evening I slipped back to Omaru Bay, and was pleasantly surprised to find Ralph Sewell's *Ripple* anchored in the shelter of the point. After a hail which resulted in an invitation aboard, Ralph and I spent a very pleasant couple of hours swapping yarns.

I slept well that night, tucked away in the shallows with my tent up. The airbed and sleeping bag in this environment are better than the best hotel bed I've ever found.

By seven the next morning we were off out into the stream and gone, heading for Graveyard Point in Awaawaroa Bay. This spot is lovely — shallow for a long way out, which keeps the big boats away, with a near-perfect beach inaccessible by road, a couple of pohutukawa trees which provide just the right amount of shade, and a grassy bank that is grazed enough to give a springy cushion under your towel when lying in the shade. This is my personal favourite in a place with many near-perfect anchorages.

I lay in the shade for most of the day, reading and dozing, swimming and beachcombing — just what the pamphlets on tropical holidays depict! In my case, the whole boat hadn't cost what an overseas holiday would, and she'd be capable of taking me on as many more holidays as I wanted. In fact, on summing up, I reckon it is cheaper to go cruising in *Houdini* than it is to stay at home!

However, the next morning's marine radio forecast told me that we were likely to have '20–25 knot winds gusting 30, with rough seas'. I figure that if you have a foot of boat length for each knot of wind speed then you'll be about right — except for extremes, maybe! On this basis, I was only about halfway to okay, with a boat a bit over thirteen feet long.

Like everyone who has to work for a living the spectre of having to be on time at the office was looming up, so after looking at the options — several safe harbours to leeward, good visibility, good cellphone coverage and a charged-up battery (yes, I had the 'batphone' aboard; I try to pretend it's not there, and although it is not a 'marine' service, it keeps me in touch with the family) — I thought long and hard about how the boat and I had performed to date.

After all that thinking — a bit hard when on holiday — I figured it wouldn't hurt to have a look outside the bay, and if the weather looked to be more than we could easily cope with, then I could duck back in. By 8.30 a.m. we were out and into it. Although the front that was causing all the fuss was not forecast to come through until early afternoon, the waves were a good metre high, and there was enough breeze to make the backs of them streaky with foam. But *Houdini* was steady enough; she was still able to tack, was making fair progress about 50 degrees

Houdini waiting for the change of tide in the very upper reaches of the Waitemata Harbour. Although very calm in here the reefed sail suggests it's blowing hard 'outside'.

off the wind, and was not taking any spray on board. We stood on, hoping to be within range of home by the next day.

About halfway across to Maraetai Beach, the wind came up a little more — 25 knots on my little hand-held meter. I didn't want any more breeze, but the boat still plugged along, seemingly unperturbed by the fuss. I had an eye on Whakakaiwhara Point off my lee bow, at the mouth of the river up to Clevedon, as a possible bolthole, or simply straight up the beach at Maraetai — shoal-draft boats have options available to them that the big boys can't use without real risk.

As the moored yachts came up on our bow at Maraetai I considered how we were going. I looked at the chart and was pleased to note that

Houdini had made good about 45° off the wind with the aid of the tide, and slightly better than 3.5 knots! Not only that, she'd been self-steering with the aid of a shock cord on the tiller for part of the leg, so we went about to see if we could make Kennedy Point, four miles away on the Waiheke side of the strait.

When one is so close to the water the patterns and shapes of the waves are much easier to see, and I soon determined that we were doing much better out in the deeper water. The waves were further apart and not so steep, and while the wind was perhaps stronger, it did not change in direction or strength as quickly. In fact, halfway across I was sufficiently comfortable to put the shock cord back on the tiller and pour a cup of tea from the thermos. Morning tea was only chocolate biscuits and an apple, a cup of tea and a muesli bar, but in those surroundings it was a feast!

While I was relaxing in my comfortable seat in the sternsheets a fairly big keeler appeared on the horizon, its spinnaker identifying it as a colleague and her husband's boat. They later told me they'd spotted my boat, recognised the design and guessed it was me, being the only

fool silly enough to be out in such weather in such a small boat! Another keeler thrashed past to leeward, driving upwind. I was surprised and pleased to see that we were, in fact, pointing as high, and that I was probably drier than the heavily rain-coated crew huddled in the cockpit of the bigger boat. Our three and a bit knots soon saw us trailing, but to some extent it put things into context. I was dry and comfortable, making good progress, and would be safely at anchor by the end of the day. What more could anyone want?

We managed to point rather higher than I thought on this board, making our landfall on Park Point, the south-western tip of Waiheke, for a well-earned lunch break.

With time pressing on we set off again. I had planned to go around the city side of Motuihe on my way to Islington Bay, but as we came up over the underwater ledge on the western edge of the Sergeant Channel, the long-forecasted front became very evident. Until then there had been a lot of cloud about, but it was well broken up and penetrated by many shafts of sunlight — lovely if a bit gloomy — whereas what was

Approaching Park Point, Waiheke Island. Still choppy but nowhere as bad as it had been further out and much, much nicer than half an hour later!

rapidly looming up was not the sort of thing I wished to be out in, no matter how big the boat. In the first smooth patch of water that came along between the steep cresting waves, I put *Houdini* about and bolted for shelter back on Waiheke to wait for the storm to abate.

As the sun set I poked the boat's nose out into the channel and found that the conditions had moderated, allowing a comfortable ride across to Islington Bay. It was dark as we rounded Emu Point, and the riding and cabin lights of the many anchored yachts made the bay look like fairyland. It was almost flat calm as we drifted up to the head of the bay and dropped the 9 kg fisherman's anchor that allows me to sleep in peace. It looks out of place on such a small boat, but I believe that small anchors are a waste of time, even on little boats, as they do not have the weight to put their points into the bottom.

It took only a few moments to open up the galley box and boil the kettle. By then my big kerosene riding light was set, the boom tent was up, and my airbed was inflated. After a light supper and a quiet read, I was ready for my comfortable bed and a good night's sleep. I wouldn't have swapped my situation in this tiny cruiser for a bunk in any of the much more pretentious craft further out in the bay!

Next morning was 'going home' day, always a sad one, but I was full of pride and reflection on a job well done. It had been about fifteen months since I had sat down at the drawing board to design the smallest serious cruiser, one with a comfortable bed, capable of coping with the weather on coastal passages, and able to make a good distance on each day at sea. *Houdini* had proven to be just about everything I'd designed her for.

With yet another gale warning forecast I left early to make it across to the weather shore to be out of the worst of it, should the dire warnings prove correct. A long tack across the four or five miles to the fortified volcanic cone of North Head at the harbour mouth and I was into more sheltered waters. It is always a good sail from there up past the Naval Base and the commercial part of the port. Auckland is a pretty city when seen from the water, and I enjoy looking at the architecture from a point of view not so closely involved in the rat race.

I made a quick stop at the Auckland Maritime Museum to visit a young friend; she and I had corresponded as she designed her own rowing boat, and I was looking forward to seeing *Gildor* in the flesh. Typical Lucy though — she was off in the little brigantine *Breeze* for a few days, so I was soon out on the water again.

Under the harbour bridge we went, with the tide helping us along, and the by now very stiff breeze kicking up a big steep sea. We bounced and splashed our way up past Kauri Point to Island Bay, our departure point a week earlier. Frank and Marcus came down to greet me, and together they helped me unload the boat and heave *Houdini* back onto the trailer.

All told it had been a good cruise — remarkably so considering the new and untried design and some positively rotten weather. I'd really enjoyed the week out, back again in waters that I'd sailed during my boyhood, visiting friends I've known for years, but with lots of time on my own for reflection and a peaceful recharging of my batteries. *Houdini* started her career well, with only a little modification needed to her plans for future builders.

I was able to conclude that the worst day sailing is better than the best day at work!

Design Nine: *Navigator*

*Dual Personalities — a club-level race trainer and
a very capable coastal cruiser*

The original Navigator *emerging from the garage for the first time. Although the structure looks complex, she was very easy to build.*

A couple of years ago I was approached by Tim Ridge of Boat Books in Westhaven Marina. A very experienced mariner, Tim wished to see a largish sailing dinghy that would be a useful family boat, seaworthy enough to go camp cruising in relatively open waters, light enough to be rigged and handled by one person for an afternoon sail in the bay, and quick enough to be competitive in a mixed club fleet. The idea was to build up a fleet of trainers that could provide a school for the club's teenagers.

Several design proposals were drawn and much discussion ensued. My existing *Rogue* design was chosen as a starting point and the parameters set. *Rogue*'s classical lapstrake plywood construction was retained, with a slight increase in length to 4.5 m (14 ft 9 in) and the beam increased to 1.8 m (just under 6 ft) to put

the crew weight out where it was needed. This also gave the boat a great deal more room, as well as making for a drier boat at speed. I called the design *Navigator*, a reference to Tim's professional background.

The wide side-decks are beneficial for both crew comfort and boat safety. In fact, tests show that she has to be heeled to almost 90° before she swamps. My 'class rules' prohibit toe straps, or sailing with any significant part of the body outside the gunwale; in this way Mum and Dad can be as competitive as the super-fit 'young guns'. The interior is laid out to give the crew an environment similar to a small high-performance keeler. One of the aims was to familiarise young crews with the systems and handling of the flat-out racers, so *Navigator* in its sloop-rigged version may have all of the controls and gear (in appropriately smaller sizes, of course) of the bigger boats.

There is a big watertight locker under the foredeck, large enough for a couple of sleeping bags and a pile of clothes. This, plus the storage afforded in the other buoyancy tanks (accessed through watertight screw inspection ports), enables a considerable amount of gear to be stowed when heading out on a cruise.

Crew weight was set at 170 kg, about two 'big people', or Mum, Dad and a child or two, or three teens. This dictated a sail area of 12.6 sq m (136 sq ft) plus a conventional spinnaker — enough sail area to be frisky in fresh weather without dropping an inexperienced crew in the tide too often.

My friend Bob Jenner sold his *Rogue* at about this time, and he turned up at my place wondering what to build next. He wanted to travel further along the coast, and explore further than he'd been able to go in his previous boat. We sorted through my stack of drawings until *Navigator* came to the top. He liked the boat but not the rig, so over several visits we developed the yawl rig based on free-standing masts and a sprit-boomed standing lug main, similar to *Rogue*'s, a rig which had him somewhat bluffed

when he first started out, but which he came to appreciate more and more as time went on.

The yawl rig, while of similar area, can be carried in much higher winds due to its low centre of effort, and the spread-out sail plan enables the boat to be trimmed for self-steering on many points of sail. If things get a bit brisk, even with two reefs in the main, she still balances under jib and mizzen, and will sail to windward with only these sails set.

Another advantage of the two-masted rig is that the boat will, with the mizzen sheeted hard and the other sails luffing, lie fairly calmly head to wind while reefs are tied in or whatever; this is a boon for people who are worried that their boat has no brakes!

Bob's *Navigator* was named *Ddraigg*, Welsh for dragon, and he and I covered 60 miles over a three-day weekend in weather ranging from greasy calm to 25 knots plus, and apart from a few exhilarating moments when surfing, we had a pretty easy ride. She covers at least as much water as the smaller keelers, and regularly surprises everybody by averaging six knots plus for hours at a time — a good cruising speed. Being the lightweight shallow-draft boat she is, she gets into corners denied to the bigger boats.

Steve Laker's 4.5 m Amy, *a cabin version of* Navigator, *on the beach at Kohimarama. Her proportions make it hard to pick the fact that she's only 4.5 m long.*

Matthew Barrie and I built this Navigator *for the gate prizewinner at the New Zealand Boat Show one year. The winner had come to the show hoping to find a sailing dinghy suitable for his young family — didn't he do well! Here, Matthew is preparing the bottom for fibreglassing.*

We've raced her, and the year Bob launched her we won all three races in our class at the Shoal Bay regatta. She's devastating in light airs; Bob has always been of the opinion that I tend to put too much sail on my boats, so I'd drawn the rig a little larger than I thought would be optimum, in anticipation of him cutting it down a little, but he bought it as it was! I was very pleased when it turned out to be about right when sailed two-handed, and easily handled by her owner when single-handed.

Building *Navigator* is simple. The 'system' is the same as that for *Rogue* and the other lapstrake-sided boats intended for beginners to boatbuilding. However, I note that Matthew, an experienced boatbuilder, who has now built three *Navigators* for my clients, seems to really enjoy building these boats, so there is something there for the skilled worker as well.

NZ Boating World magazine did a test on *Navigator*, sailing Bob Jenner's yawl for a couple of hours one murky fresh winter's day. The 'tester' was Andrew Mitchell, who had just returned from coaching the New Zealand Youth yachting team at an international regatta in Italy.

He enjoyed the sail a lot, finding the combination of traditional rig and styling with modern materials and underwater shapes both attractive and practical. He wrote a glowing report which sold about 40 sets of plans in the months following publication, and as new boats hit the water the interest they created sold more and more plans.

There are quite a number of *Navigators* about now. One or two are the terror of their club's all-comers dinghy fleets, and many are doing a wonderful job of extending their owners' horizons and taking them cruising further and further into new waters.

One thing that has surprised me — about nine out of ten are being built with the yawl rig! I would have thought that a two-masted cruiser only 4.5 m long would have been a bit too specialised for today's yachtie, but the cruising *Navigator*'s popularity makes me think there is hope yet in these times of politicians, futures brokers, used-car salesmen and other high profile fakers.

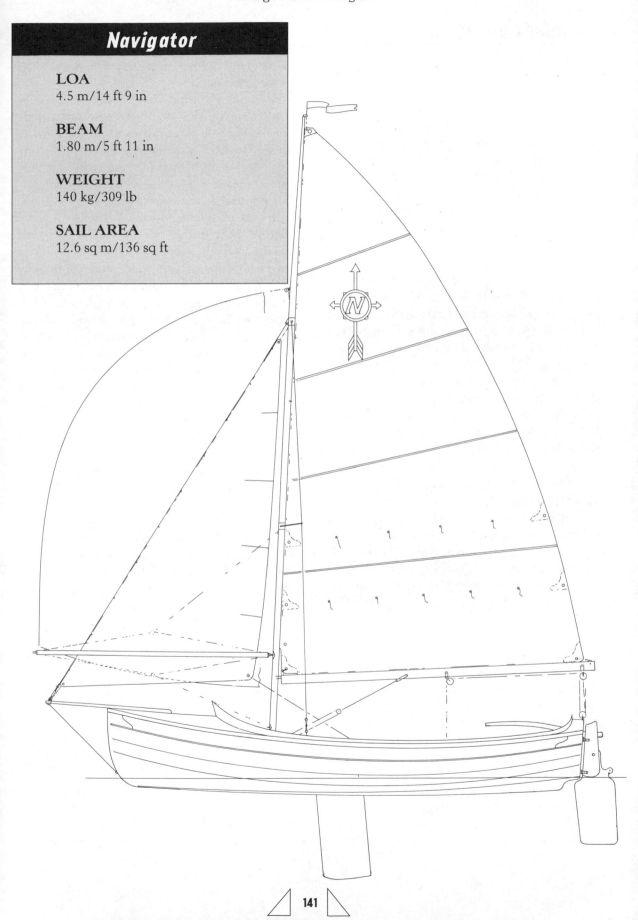

Navigator

LOA
4.5 m/14 ft 9 in

BEAM
1.80 m/5 ft 11 in

WEIGHT
140 kg/309 lb

SAIL AREA
12.6 sq m/136 sq ft

Another Little Gem

Bob Jenner and I had been friends for many years, and I cannot recall any disputes or hard words between us, but as I pulled the tiller over to put *Ddraigg*'s bow down the 1.5 m wave and sheeted the big main on hard, there was a tone to his voice that suggested that perhaps I was pushing my luck! Still, he sat amidships as instructed, and the little 4.5 m yawl took off down the wave with a burst of acceleration and a cloud of spray. I suspect that his fingerprints might still be imprinted in the varnish on top of the centrecase!

We'd been out in his *Navigator* design for a day at this point, leaving Narrow Neck Beach, on Auckland's North Shore, with the idea that we might sail around Waiheke Island over the long weekend. Our first day out had been a fairly gentle sail past the volcanic cone of Rangitoto, through the Motuihe Channel, and along the northern side of Waiheke, poking our noses into the bays and coves as we went.

Garden Cove was our first stop for the night. Bob is a very experienced camper and a keen sailor, but this was the first time he'd combined the two disciplines. Our night ashore was comfortable and contented, with much discussion about the possibilities of this 'new' sport.

Next morning saw us ready to go before the weather was, and it took a little sculling to see us out through the narrow entrance to the little harbour. As the first ruffled streaks appeared upon the water, *Ddraigg* ghosted along, giving us a close-up view of the reefs and cliffs that form the coast here.

Red-blossomed pohutukawa, golden sandstone cliffs, and the undulating flat blades of kelp cannot be appreciated from afar from a larger yacht; the activities of the birdlife, and the fish among the rocks and sandy patches made our slow progress fascinating. We enjoyed the gentle drift, but were pleased when the breeze filled in about mid-morning as we had a fair distance to cover that day.

We sailed past Gannet Rock, watched by the birds commuting to and from their fishing grounds, and out around Thumb Point, standing well off to avoid the swirling current set up by the underwater ridge that runs out from the point. It must have been the last of the ebb tide as the wind did not create any significant chop as we sailed by.

Although the wind was gradually increasing, it was also swinging round and heading us as we worked our way past the measured mile markers above Hooks Bay, so getting an accurate time was not possible. It pays to take every opportunity to measure speed so one can judge the distance run when passage-making.

It was while entering the Waiheke Channel that we went surfing. Ocean swells, set up on the shallows and being slowed and steepened by the now fresh southerly wind, made for troughs high enough that we couldn't see out of them. A steep following sea some 1.5 m high and a fresh headwind were too much temptation for me on the helm; I sheeted the big main on hard and it was all on! Bob is a fairly conservative sailor and had never been surfing in a boat of this size, so the first couple of runs, with gunwale-high bow waves and incredible acceleration, were a bit hard on his nerves!

He got over it, and lunch on the beach settled him even further. However, I've since noticed a distinct expression of concern on his face whenever we are running down big seas. It must be my imagination!

We camped at my favourite spot in Awaawaroa Bay that night, a peaceful night with the boat left afloat rather than beached. As I nodded off to sleep an old saying drifted — appropriately for us — into mind: 'the size of your adventure is increased by the smallness of your boat'!

The next morning, I attempted to make up for the previous day's sins by being the one who swam out to fetch the boat in for loading, and we were soon away. From Waiheke's Bottom End to Auckland is a sail that always fills me with mixed emotions. One starts with a feeling of wilderness, which gradually gives way to the crowds of houses and bustle of the commercial port, and — on this day — the crowds of racing yachts, dashing about like flocks of white sheep. Although much bigger and more pretentious than *Ddraigg*, they were not in the same league as us — 'we' were returning from a 'voyage'!

Frank Bailey and I sailing Ddraigg in the Rangitoto Channel. With one reef in the main, we're still moving along at a real clip. Photo: Andrew Mitchell

Design Ten: Six-metre Whaler

A sailing trainer with real potential for fun

Fred Jeanes is a man of many parts (some would say too many). All joking aside, he has some laudable ambitions about educating and changing the attitudes of some of the 'at risk' young people in our society, especially in Tauranga where he lives. When I met him he was at the stage of asking around to see if what he proposed was workable.

A critical part of Fred's programme was an open boat which would provide a teamwork environment as well as transport on the vast area of shallow tidal waters that spreads from Tauranga north to Katikati. Enquiries as to the cost of a new naval whaler were not at all reassuring; the figures varied a lot, but they were all far above the sort of budget that Fred had in mind.

This is where I came in. After discussions I suggested that, for around $5000 plus volunteer labour, we could produce a boat that would carry six trainees plus an instructor in semi-sheltered and coastal waters. She would pull six oars, have provision for an auxiliary outboard motor, and would sail well enough to make real progress under sail.

When sailing there would be a job for each of the crew, seeing to the port and starboard jib sheets, mainsheet, centreboard, mizzen, and helm; the instructor could sit back and instruct; she would have enough buoyancy to float at a useful height if swamped, and would be light enough for the normal crew to manhandle up a beach to set up camp for the night.

Best of all, at a whisker under six metres long, she would not be required to be built under the expensive system of government-controlled survey. We would build her as soundly and as safely as if she were, but the costs involved in a survey system intended for full-sized ships were outside the budget's capacity.

In order to stay within the popular image of a sail training boat, we needed to have a vessel that looked the part. Most non-boating people see the Naval Whaler as being the ideal training boat for these outdoor education centres — but they don't have to handle or maintain them!

I drew up a double-ended open boat which became the Six-metre Whaler, with two masts and a cheeky sheer. The design was strongly based on the very successful yawl-rigged version of the *Navigator* design.

She has a pair of straight alloy tubes as masts, with the same sprit booms and similar proportions to *Navigator* in the sails. All spars fit inside the length of the boat for storage, or to reduce windage for rowing, and there is a steel centreboard to make her a little more steady. (Truth is, I wanted to be able to go down to Tauranga and sail one of the boats from time to time, and the stabilising effect of the heavy board would make her easily handled by a crew of two. She is even more highly powered than *Navigator*, and I wanted to see what we could do to the local trailer yacht group!) The structure was kept simple so that once patterns were made while building the first one, others could be built quickly and easily using conscripted labour.

Fred approved the final lines, and gave me the go ahead to produce the construction drawings. The plans are now finished, but Fred is unfortunately stalled. There is considerable support in the city for the project, but money is a little less forthcoming. Fred's team have done a trial run in a sea scout cutter, demonstrating the advantages of the Six-metre Whaler for the waters in which they will be working.

Meanwhile, the whaler has attracted a lot of attention, and two boats are on their way. I've sold a couple of sets of plans to people who are going to use them as dayboats for family holidays. One is badgering me for drawings that convert her into a cabin boat with minimum but sheltered accommodation — a thought that intrigues me!

Six-metre Whaler

LOA
5.99 m/19 ft 8 in

BEAM
2.14 m/7 ft 0 in

DRAUGHT
Board up 0.26 m/0 ft 10 in
Board down 1.35 m/4 ft 5 in

WEIGHT
(including centre board)
300 kg/661 lb

SAIL AREA
16.25 sq m/175 sq ft

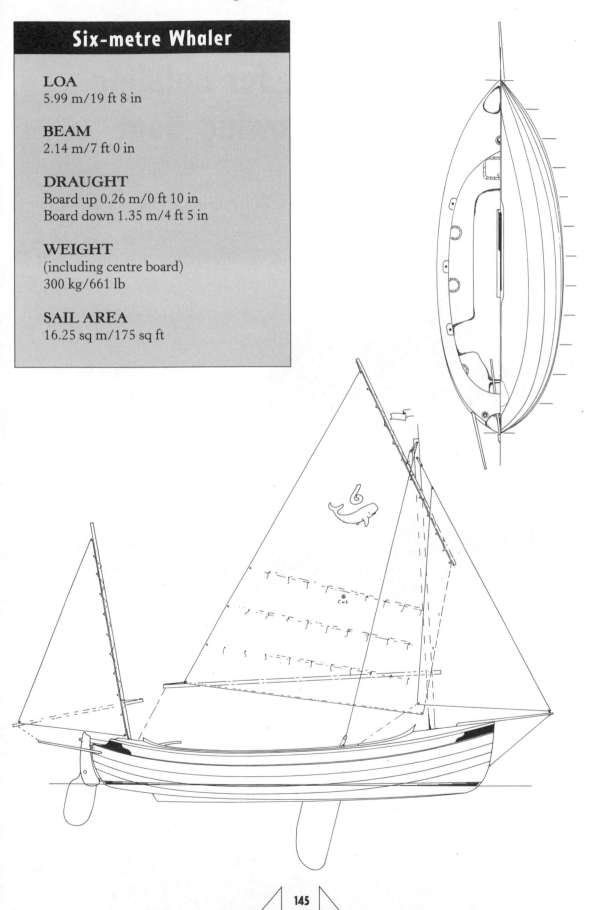

Considerations for Building a Cruising Rowing Boat

A case of boats, backs, butts and blisters

It was this boat, my Light Dory, built for a seven-day cruise on the Kaipara Harbour, that sorely tried both the boat and the rower, but taught me a great deal about cruising under oars.

Cruising in a rowing boat is a lot like tramping afloat: a way of visiting the coast's less populated spots, and a way that returns rewards far beyond the modest resources required. It offers the rower an unequalled experience of the shore, and there are many places, a large proportion of which are inaccessible to bigger craft, that are as yet quite undisturbed.

In selecting, building, or — for the brave — designing a touring rowing boat, serious consideration must be given to the ergonomics of the boat. By that I mean that the boat's seat, footrests and rowlock layout should suit the build, suppleness and strength of the rower. For example, for someone a little older and perhaps a little stiff in the hips or lower back, it helps to have the seat at least 200 mm higher than the heels of the footrests. A fitter, more supple person might be able to tolerate a lower seat, and gain a little stability from the lower centre of gravity, but I use the 200 mm as a standard height for the touring boats that are my main interest.

In a boat intended for open water, the footrests should be at least 400 mm apart. This applies to sliding-seat boats as well as the fixed-seat boats you see here. The habit of putting the footrests close together is, to my way of thinking, a regrettable one that comes from the skiffs or shells that are too skinny to do otherwise; it is a major handicap when the boat is being rolled about in a seaway. With the feet wider apart, the well-braced rower can control the upper body position, and will find it much easier to keep rowing when the going gets rough.

Positioning the rowlocks is also very important. After much experimentation I now place the rowlock pin 350 mm aft of the rear edge of the seat, and 220 mm higher. To set the relationship between the seat, rowlocks and footrests, the oars should be roughly straight across when the elbows are just ahead of the ribs, with the rower sitting dead-straight upright and feet firmly on the footrests. Just a point: rowing without footrests is not much fun — it leads to sore backs, sore butts, and not much progress for the energy expended!

A boat intended for open water rowing, recreational or racing, depends on length for its speed. Typically the successful designs are between 4.5 m and 5.75 m long, with a beam of between 1.25 m and 1.35 m at the gunwale to provide a good spread between the rowlocks. She will be quite narrow at the waterline to reduce drag, and will be as light as possible, consistent with the technology available to the builder and the manner in which she will be used (or abused).

The uninitiated will find rowing one of these craft somewhat like the first time riding a 'two wheeler', but — like the bike — it will not be long before you wonder how you could ever have felt unsafe. In fact, with practice, some of these boats can be extraordinarily seaworthy, as long as the rower has the energy to keep them moving at the right angle to the waves. Note that we are talking mainly about the non-outrigger fixed-seat boats which form the majority of the recreational fleet.

I can hear the 'old salts' muttering about the lightweight boat not 'carrying its way' (gliding on between oar strokes), and can assure the doubters that many of these very light craft will keep moving for a long time while the rower rests. They behave very differently from the old clinker dinghy that Granddad used to keep down by the beach, and will generally outperform the common general-purpose dinghy of yesteryear, even in really adverse conditions.

Seaworthiness is a very important consideration in the design of an open water rowing boat, as the rower cannot be expected to outrun a squall in the same way that a powerboat can. We designers must provide our clients with boats that will cope with the worst that one might encounter when crossing the mouth of a big estuary, for example, wind against tide can create sea conditions that are out of all proportion to expectations. The wide flared sides that give the rowlocks their spread, combined with the strong sheer that most of the traditionally styled craft have, help the boat ride over breaking crests, while the narrow waterlines and fine bows help it drive through head seas in a manner impossible with other kinds of boat.

As mentioned above, the ergonomics of the boat are critical. A movement that will be repeated thousands of times must be both effortless and truly comfortable. To give you an idea, when I am cruising I like to row for two hours in the morning and two hours during the calm of the evening. At my favoured pace of 25 strokes per minute this adds up to 6000 strokes per day, far too many repetitions to tolerate even the slightest discomfort.

In terms of movement, the exaggerated pendulum action of the torso, seen in competition craft such as surfboats, has no place in long-distance cruising. Not only does it burn energy far too quickly, but on a long trip it stresses the body more than is desirable.

The calculation of oar length for one of my boats is based on a movement at the hands of only 700 mm. The theoretical cruising speed of the boat is then worked out by taking the square root of the boat's waterline length in feet (imperial units of measurement are good for some things) and multiplying the result by a figure between 1 and 1.4. This figure is arrived at by an analysis of such things as the boat's displacement-to-length ratio, beam-to-length ratio, and the entry and exit angles of the waterflow. All very scientific, but what it really means is that a short fat heavy boat will be close to '1' while a really long light slippery boat will be at the other end of the scale.

All this jiggery-pokery tells me how fast the water will be moving in relation to the boat, and by applying a slippage factor appropriate to the oar type, about ten to twelve percent for the narrow blades I use, I can work out the length of shaft needed to move the blade at the right speed when the handle is stroked through 700 mm 25 times per minute! There are other factors and variables, but this will give you a fair idea of the process, and will get you within adjustment range of the right figure in any normal boat.

Where the handle should be as you pull the oar through is a subject of some contention among rowers. Those who have experience in single sculls (the 'flying toothpick' school of rowing), tend to prefer the grips to be cross-handed or overlapping, even on the pull. I cannot see an advantage to this, and prefer to have the handles half-overlapped on the recovery only; this leaves the handles far enough apart on the pull to get my thumbs around the inboard ends of the handles to give the palms a rest, and try to reduce the chance of blistering.

Another way to help this is to use one of the specialist adhesive bandages sold to the long-distance running fraternity. I use the Spenco™ brand, applied over the entire palm and up to the second joint of the first two fingers only. I make sure that the movement of the hand is perfectly free, and replace it every two days when cruising; the stuff sticks like the proverbial to a shovel, and breathes well, so there is no problem with the skin underneath. I've tried all of the normal cures, including meths, gloves, and vitamin E cream (I drew the line at urine!), but have found the sticking plaster the best yet.

The wondrous rowlocks seen on the rowing club skiffs and shells are a bit too expensive and far too complex for my knockabout use. I stick to a conventional bronze rowlock from a local foundry, and set the pitch by carving a small 'flat' in the underside of the oar handle where the thumb grips it. When positioned correctly this will make the oar fit the hand only at the correct angle, and will keep the blade at a constant angle in relation to the hand.

Since I had the misfortune to injure my hands

on a woodworking machine, I have taken the shaping a little further and made a similar small flat on the top side for the first two fingers, which greatly improves the grip of my no longer strong hands. Friends who have rowed with these oars have, in many cases, gone home and done the same to their oars.

Note that the 'catch angle' or 'pitch' of the blade should be about 15° positive on these boats. That means that the blade is angled 15° to dig into the water as you pull on it. Not only does this prevent the blade from popping back out of the water as you heave on it, but it tends to stabilise the boat as well.

Good oars are essential; they make or break the whole project. My efforts to build good ones from scratch were successful, but I conceived a serious dislike of what was for me a tedious job, one that took too long while my new boat sat waiting to get out on the water. Nor was I prepared to pay through the nose for some craftsman to charge me the earth for 'specials' (you will have gathered that I don't think much of the store-bought models!) so now I buy standard seconds in local pine and doctor them myself.

I cut the blade down to 110 mm wide at the tip, tapering to 90 mm where it begins to curve into the shaft. The loom (the section of the oar's shaft from the blade inboard end, or 'neck', to the leather, where the oar passes through the rowlock) is also tapered, leaving it close to the original dimensions fore and aft (90° to the blade), but tapering from round where it leaves the leather to only 30 mm at the neck; this is rounded off to a nice oval shape.

A round-mouthed spokeshave is used to thin the blades to about 6 mm at the blade edges, and 9 mm at the centre of the blade, with a full-thickness rib down the centre. This rib can be carried out to the end of the blade, and the tip scalloped to form a strong pointed end to the oar. It doesn't help the performance but it looks appealing, and gives you something to push the boat off a jetty with without breaking the oar's blade!

My oars have very long leathers made of leather, of course. But I do secure them with contact cement before stitching them, and then coat the thread with glue which prevents the stitching from being worn away. There are no 'buttons' or lugs to locate the oar, but rather a lanyard from the rowlock throat to the neck of the oar. There is a tent rope adjuster on each one should the length need changing; this enables the 'gear ratio' to be changed by moving the handles inboard and rowing cross-handed into a headwind, or slid out a little if rowing leisurely downwind. This system also serves to secure the rowlocks into the boat, meaning that the oars stay with the boat in case of a mishap!

To test an oar's balance, sit in the boat with the oar in position, resting the hand on the handle with the shaft straight across the boat; the tip of the blade should only just kiss the water with the forearm completely relaxed. If the oar is still heavy in the blade after being 'operated' on, it might pay to try a lead collar on the shaft up by the handle.

It can take a while to get used to the feel of a well-balanced pair of 'sticks' but it's well worth the effort; they burn a lot less energy than the overweight 'clubs' that most people use.

Earlier when I described the dimensions of the blades, my ears burned a bit. I know that the big spoon blade is the accepted norm for performance boats, and for short distances in flat water; when swung by trained athletes this is certainly the case. However, if you look at the history of rowing boats used in open water, you will notice that they almost invariably had very narrow blades and allowed extra length for slippage.

These slim blades were not only less prone to damage, they had a lot less windage. The old fishermen knew that having to twist the wrists 1500 times an hour for half a day at a time to feather those big blades was just not practical. Should you not agree, think of me when you are sitting out on the water, your forearms near-rigid with tendonitis, carpal tunnel syndrome, or RSI.

Still on the subject of bodily comfort, the seat is very important. Sitting at 200 mm high, the knees are too high for the thigh muscles to help the buttocks cushion the pelvis, and the rocking action can make for a very sore tail! I have a piece of soft closed-cell foam over which I have a shaggy sheepskin — luxury, no matter what one might or might not wear. Oh yes — while in this area of the anatomy, I use the bailer! There are times when the shore is too far off, and standing close enough to the gunwale to get 'it' reliably over the side is impossible in these tippy boats.

Food and drink are an important part of your cruise. This is the fuel which propels your craft, and it is possible to run out while under way. On one of my longer cruises, I kept a close record of my food intake, and found that I was consuming around 4500 calories per day, rowing on average five hours, plus walking and exploring. I lost fat and put on muscle at about the same rate, so my weight stayed constant.

I find that some slim people have to stop and eat every hour or so, and everyone should drink at least that often as there is a real danger of dehydration. To stop more often, or for longer than a few minutes, means having to go through the warm-up again as the muscles will have cooled and stiffened, a painful process if you've been at it for a few hours.

For snacks on the move, the trampers' favourites, such as 'scroggin' and muesli bars, are very useful. There are special dietary 'boosters' for endurance athletes, but I don't like the taste much and they cost an awful lot. My drink of choice is a simple cup of tea from the thermos.

I've covered the subject of food and cooking in a previous chapter. Suffice to say that I put a fair bit of time and effort into cooking good food when camping, and see no need to exist on the frugal fare that some use when away from home. In fact, one of the advantages of cruising rowing boats, as opposed to tramping, is that one's epicurean preferences are not restricted by one's ability to carry the weight involved.

Imagine easing your boat up the estuary, keeping an eye out for a spot just big enough for your tent, the roar of the city just audible as a counterpoint to the near-silence of your craft — rowing is almost the perfect way to travel. When you only have a few days, a lot of pressure in your day-to-day life, and little income to spare on recreation, open water cruising in a rowing boat can be every bit as good for the soul as it is for the body.

Design Eleven: *Seagull*

A simple recreational rowing boat

A beautifully finished example of Seagull, *clearly illustrating the wide flat bottom which gives the craft its envied stability.*

I am very keen on recreational rowing boats. My wife Jan prefers to see where she is going in her sea kayak, but I like the feel of the bigger boat, and the way the rowing motion uses all of the body's muscles, as well as assisting the breathing as the torso rocks back and forth.

To row a boat such as *Seagull*, or any of the other rowing boat designs, is easier than walking, and surprisingly fast. Carrying a load, such as camping gear or a friend or two, makes little difference to the effort needed to move along at an easy pace, and it is a wonderfully peaceful way to explore an estuary or coastline. For those who, like Jan, wish to see where they are going, a motorcycle wing-mirror fitted to the gunwale makes a dandy 'front view mirror'! For those who wish to use an outboard motor ... why not? *Seagull* has a transom braced for a small motor, and will get more performance out of a 2 hp engine than anything else I've seen.

Recreational or sports rowing boats are often pretty easily upset, the narrow waterline beam making them very tender. In this boat though, the bottom is wide enough to allow one to (carefully) stand up within the boat, stability which, although it knocks a little off her performance, makes her a much more versatile craft, and only about three minutes in the hour slower than her more sophisticated sister *Joansa*, featured a little later on.

Seagull's sides are sewn to the bottom, awaiting the fitting of the transom. You can see the shape starting to appear.

Just idling along — Seagull at the Mahurangi regatta.

Seagull is built in the same way as *Fish Hook* and her other sisters, and in fact the long gentle curves of the proportionately slimmer rowing boat make her easier to build, in spite of the larger overall size. If you use Gaboon or Occume plywood, and take care in the building process, it should be possible to keep her weight down to about 40 kg.

Rowing a boat like this is something a lot of boaties never experience. They move incredibly easily using very little energy, and once the tech- nique of using the oars has been mastered they can be rowed for hours on end at a speed that would surprise many sailors. A good recreation- al rowing boat, unlike the delicate 'toothpick' that the competitive rowers use, can cope with quite extreme weather conditions, and with an experienced rower in charge will ride over seas that would put much larger boats at risk.

Gently stroking *Seagull* along on a quiet evening tide, up an estuary somewhere not far from here, is my idea of heaven.

Seagull

LOA
4.64 m/15 ft 3 in

BEAM
1.226 m/4 ft 0 in

WEIGHT (approx.)
42 kg/93 lb

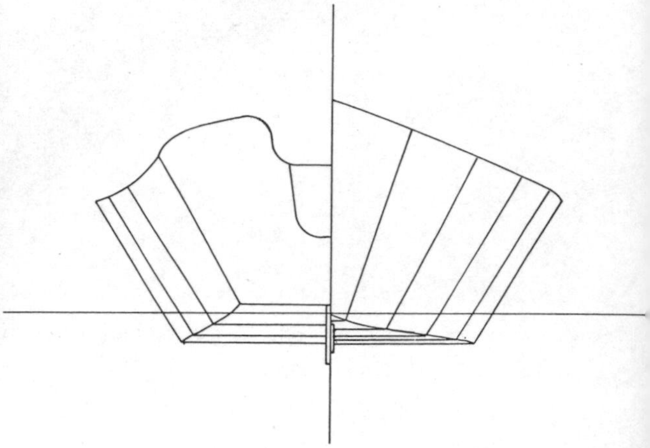

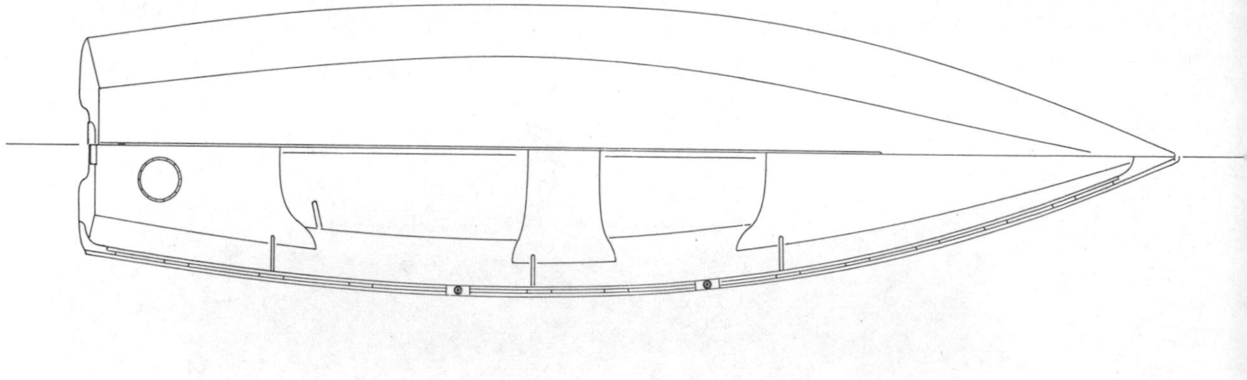

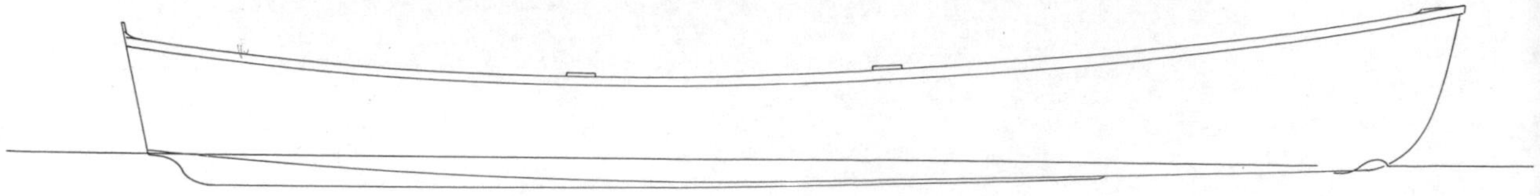

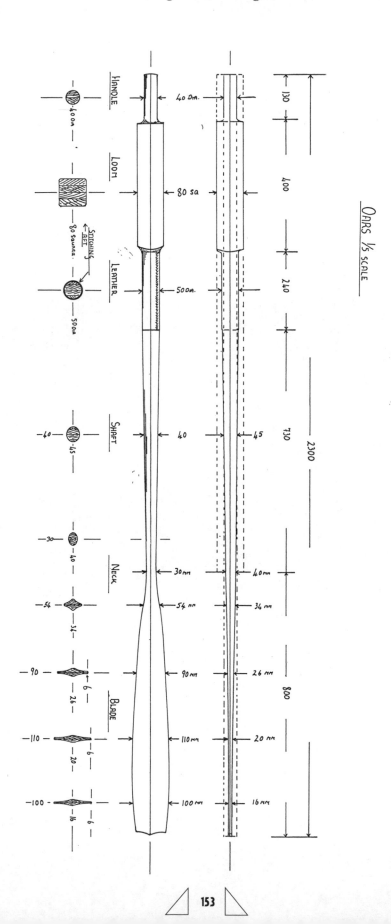

OARS ⅕ SCALE

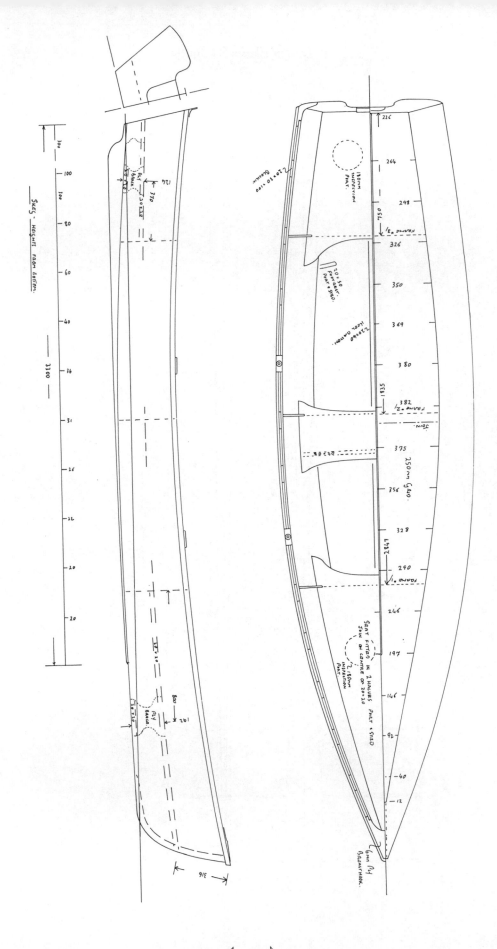

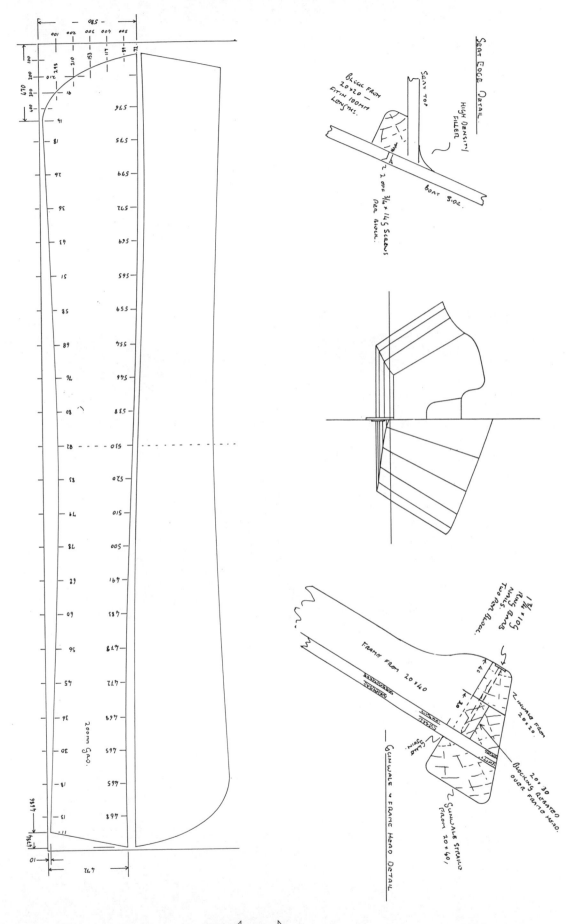

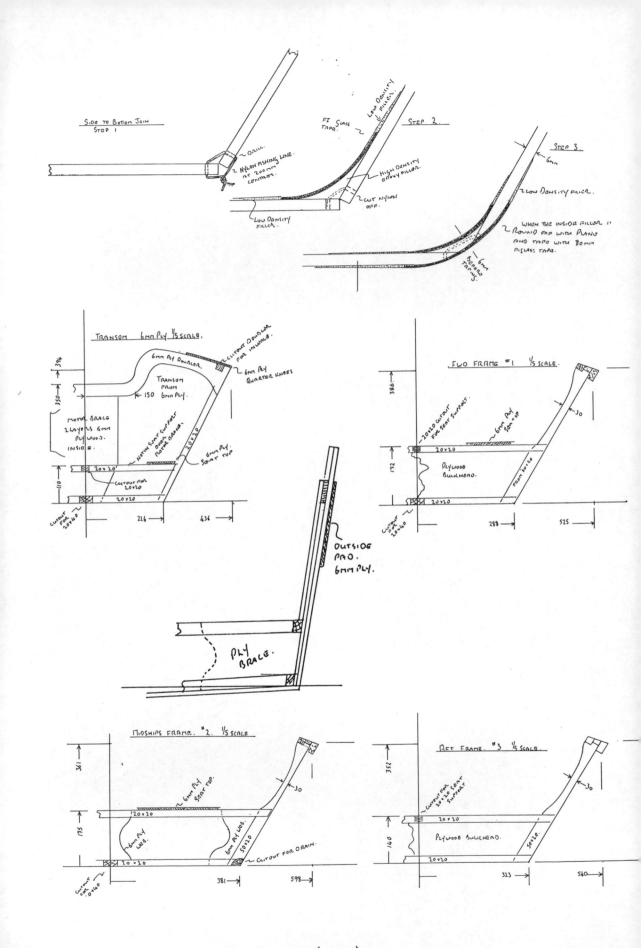

SIDE TO BOTTOM JOIN
STEP 1

DRILL.
2 NYLON FISHING LINE AT 200mm CENTRES.

LOW DENSITY FILLER.

FI GLASS TAPE.

STEP 2.

LOW DENSITY FILLER.

HIGH DENSITY EPOXY FILLER.

CUT NYLON OFF.

STEP 3.

6mm

LOW DENSITY FILLER.

WHEN THE INSIDE FILLER IS ROUND OFF WITH PLANE AND TAPE WITH 80mm FIGLASS TAPE.

6mm 2 LAYERS TAPE.

TRANSOM 6mm PLY. 1/5 SCALE.

6mm PLY DOUBLER.

CUTOUT DOUBLER FOR INWALE.

6mm PLY QUARTER KNEES.

TRANSOM FROM 6mm PLY.

150

MOTOR BRACE 2 LAYERS 6mm PLYWOOD. INSIDE.

NOTCH SEAT SUPPORT INTO MOTOR BRACE.

20×20

6mm PLY SEAT TOP.

20×20

CUTOUT FOR 20×20.

20×20

CUTOUT FOR 20×40.

216   436

394
350
110

FWD FRAME #1 1/5 SCALE.

386

20×20 CUTOUT FOR SEAT SUPPORT.

6mm PLY SEAT TOP.

20×20

PLYWOOD BULKHEAD.

FROM 20×20.

30

172

CUTOUT FOR 20×40.

20×20

288   525

OUTSIDE PAD. 6mm PLY.

PLY BRACE.

MIDSHIPS FRAME #2. 1/5 SCALE.

361
175

6mm PLY SEAT TOP.

120×20

6mm PLY WEB.

6mm PLY WEB.

50×20

30

20×20

CUTOUT FOR DRAIN.

CUTOUT FOR 20×40.

381   598

AFT FRAME #3 1/5 SCALE.

352
140

CUTOUT FOR 20×20 SEAT SUPPORT.

20×20

PLYWOOD BULKHEAD.

50×20

30

20×20

323   560

Design Twelve: Light Dory

A seaworthy rowing boat

My Light Dory all set for the next leg of the journey down the northern Wairoa River; the gear is stowed and covered, the spare oars are lashed in, and my lifejacket is ready to put on.

It suddenly dawned on me one bright October morn that, barring miracles or extraordinary effort, there would be no Christmas cruise that year. The first job was to decide on a boat. I had reservations about the shapes of some of the craft I looked at, the construction method or costs of others, and concerns about the seaworthiness of most, so DIY (design it yourself!) seemed to be the order of the day. One of the considerations, of course, was time. The way to

build boats really quickly is not to work like lightning but to build boats that are very simple and relatively small!

I'm not sure when the idea first began but the notion of a boat for 'cruising under oars' intrigued me. Among my reading research that day was an article by a guy who had been cruising in his traditional rowing dory, a book on the history of the Kaipara Harbour, and a study on a Phil Bolger-designed light dory.

I'm not sure if it was the steak or the red wine, but sleep did not come easily that night. My mind ran riot, combining bits of the day's reading with my memories of camping and visions of the endless miles of deserted beaches on the wild west coast. By two a.m. I was up again with my New Zealand atlas, looking at the Kaipara's vast and sparsely settled shoreline. Only a couple of hours north of home, the Kaipara Harbour is almost 70 miles in length, and has what is reputedly the longest shoreline of any southern hemisphere harbour; it looked like an ideal place to get away from it all.

By midday next day I had drawn up plans for a boat that would be suited to the fast tidal currents and short steep chop of this notoriously unpredictable stretch of water. The boat was similar to the Bolger light dory: a little longer, and customised both to suit my build and to cope with the worst conditions possible, conditions which I fervently hoped not to encounter.

What eventuated was a plywood dory with a tombstone transom (it's not really a dory otherwise), a nicely curved stem to force some shape into the forward sides, a strong sheer kicking up

aft to a high stern (thoughts of going surfing), with the maximum beam further aft than is traditional to make the boat run straight in heavy following seas. Added to this was a big skeg for directional stability and to balance the windage, buoyancy tanks under the seats, and a sculling notch in the transom. The boat was beginning to look very much the part.

After the usual laborious arithmetic and fiddling with moments and volumes, the sketched ideas were turned into line drawings, and the drawings into a $1/10$ scale model.

For speed of building, the dory was built using 'stitch and tape' techniques, eliminating most of the framing and all of the building jig. Basically, flat panels were cut to shape and then joined along the edges to give the desired shape. Some minimal framing carried the loads from the rowlocks and the seats, and the seats formed buoyancy tanks in case of a major disaster. The Light Dory was, in construction anyway, the immediate ancestor of the four 'simple' stitch and tape boats featured earlier.

Predetermining the shape of the panels was the difficult part. In this case I built the model

Wayne Chittenden's Emma McLeod, *a Light Dory design, at the annual Mahurangi regatta. Note the 'front view mirror' fitted to the gunwale!*

Wayne Chittenden cruising along in Emma McLeod *near Herald Island.*

over solid station frames, and, when finished, wrapped light card around the boat, traced around the edges, cut the shape out, and transferred it to graph paper, making scaling up an easy job.

Building the boat did not take long. A few hours here and there over a couple of weeks saw her ready to paint, and the total cost of about $250 did not hurt the pocket too badly.

I can carry the boat on my own, essential for a boat that will be used almost entirely single-handed; she weighs in at 42 kg dry. She rows beautifully; at a steady 24 strokes per minute on the customised 2.3 m (7 ft 6 in) oars, I can make 3.6 knots through the measured quarter mile without busting my boiler. Stable and comfortable, she tracks well, and handles better loaded than light because of the heavily rockered bottom. She cannot be driven past hull speed, even by a very strong person, but is wonderful for eating up the miles at a moderate pace.

The light weight and the narrow waterline beam combine to make the boat feel a little tender when unloaded, especially when the rower is unused to the wiles of dories. I carry about 30 litres of water in plastic jerrycans which improves her stability, and also helps the boat carry its way.

Like most fine-ended boats, she is pretty sensitive to trim. If I find she needs to be a little bow down to stop her blowing around broadside on while I try to row her to windward, or conversely, stern down for downwind, I have a ten-litre plastic container of water on a light line and toss it to the appropriate end of the boat. The line is so I can retrieve it without moving from my secure seat amidships.

A full load of camping gear and provisions makes things easy; the boat rides better down on her lines a little, and she is much more stable. Rough weather in one of these is exhilarating; the motion of the boat is very easy, especially in

My campsite near Scotts Landing on the Mahurangi Harbour. This lovely sheltered spot was 'home' for two days.

a beam sea — the rower's position is right on the centre of motion about which the boat pitches, so the oarsman doesn't get thrown about in a seaway. It is amazing to watch great foaming crests disappear under the gunwale without any fuss.

The Light Dory has proven to be not only great for recreational rowing but was ideal for its intended job, the trip under oars from Dargaville, in the Kaipara's northern reaches, to Helensville, 70 miles away in the south. The story of this voyage is far too long for inclusion here; suffice to say it was one of the best times of my life! There were a lot of personal firsts, and some real challenges. I covered about 120 nautical miles in the six days that I was on the water, seeing parts of New Zealand that I didn't know existed, and arrived at the other end of the harbour with my batteries thoroughly recharged.

This trip, undertaken because of a shortage of both time and funds, started me on a path that has seen me design a number of recreational rowing craft, cruise many parts of the coast around Auckland, do a bit of racing, and make a study of fixed-seat recreational rowing boats.

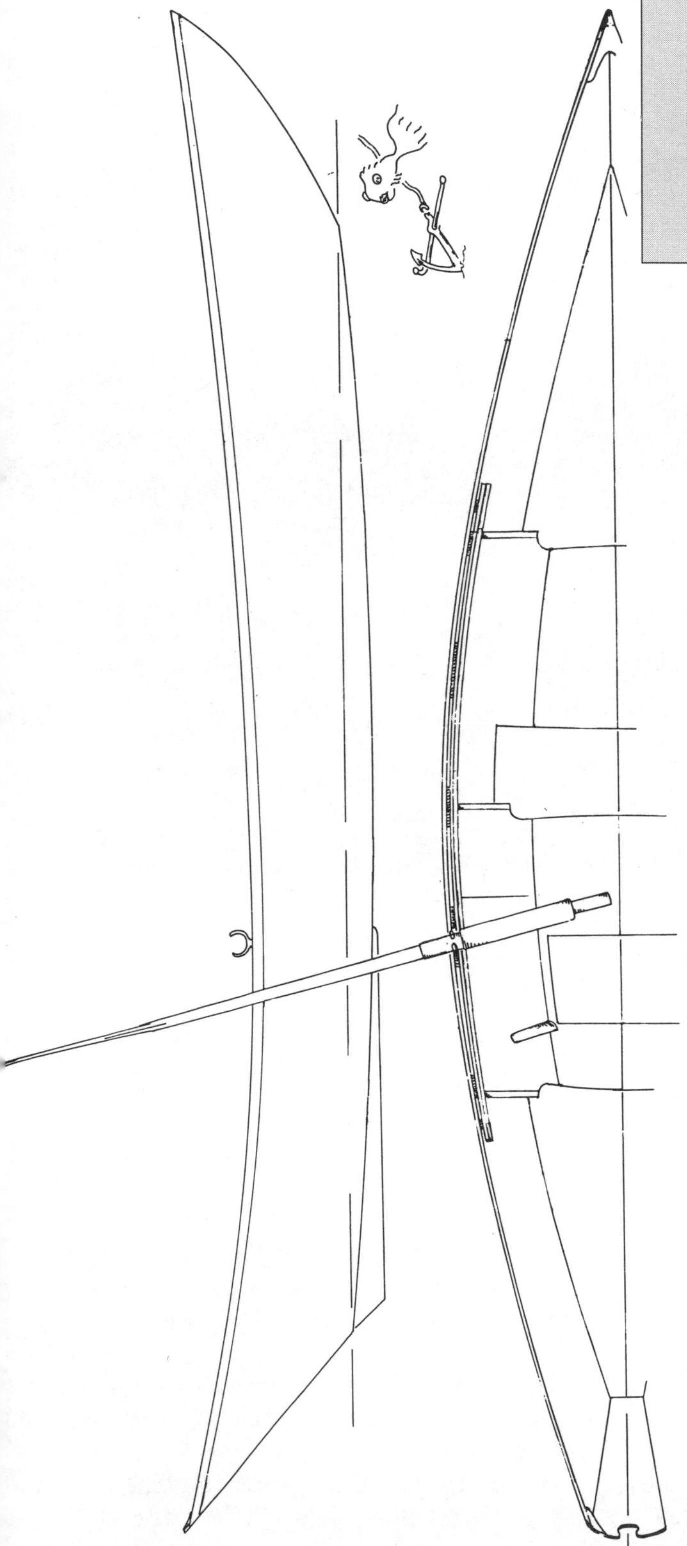

Light Dory

LOA
5.14 m/16 ft 10 in

BEAM
1.26 m/4 ft 2 in

WEIGHT (approx.)
42 kg/93 lb

Design Thirteen: *Joansa*

A sporting rowing boat

A very nicely finished Joansa. *This boat is staggeringly fast under power, even when the motor is just this little Seagull.*

A while ago I spent a season cruising in a very capable light dory I'd designed (featured in the previous chapter), the trips ranging from a week on the Kaipara Harbour to an afternoon trip up to Riverhead and back.

It's a lovely way to see the coastline, surprisingly fast in the right sort of boat, and a lot easier than shouldering a pack through the bush. A doctor friend tells me that fixed-seat rowing is one of the most beneficial and least destructive forms of exercise. (I've no doubt that exponents of other disciplines will disagree, but few dogs will bother chasing a rower!)

I sold the dory to fund the next design and building project, but I missed the ability to be afloat within moments, and the glow of well-being after a good hard row down the channel. I missed being able to take my wife and daughter out with the thermos and a picnic lunch, exploring places as dissimilar as Westhaven Marina and the upper reaches of the Puhoi estuary.

Well, the workshop was empty, so it was back to the drawing board again. I fancied one of the classic round-sided dories; the last one had been one of the better-known straight-sided types adapted to plywood and to our local conditions. With this new boat I'd a mind to try for the efficiencies of the more complex round-bilge shape while trying to stay with the dory's simple construction.

We called the boat *Joansa*, a name derived from our family — John, Jan and Sarina. (You try and do better with that lot!) There is now a Brendan, but he wasn't even a twinkle then, if you know what I mean. *Joansa* is an adaptation of a late 1800s Chamberlain dory skiff, native to the inshore fishing industry of the north-east coast of North America — smart rowers, seaworthy, simple to build and maintain, and good lookers.

My modern-day version is light enough to roof-rack on our old Toyota Starlet, fast enough to win races, and carries enough weight to take the three of us plus gear on a week's camping holiday.

Joansa proved popular. She is a very pretty craft that attracts a lot of attention, and plan sales were so good that at least two boatbuilders took them on as a 'stock boat'.

Our *Joansa* did a lot of miles for us, travelling quickly and easily; some of the mileage was done under power with our 2 hp Honda outboard, but mostly under the oars she was designed for. We loved her lots, but inevitably the time came around for another project, and it was a couple of years later that we built *Seagull* to replace her.

Joansa *being planked up. This view inside the bow shows the central spine and stem slotted over the number one frame and the light stringers.*

Frank Bailey and I with Seagull *(in the foreground) and* Joansa *(behind), stopped for lunch near the headwaters of the river at Riverhead, Auckland.*

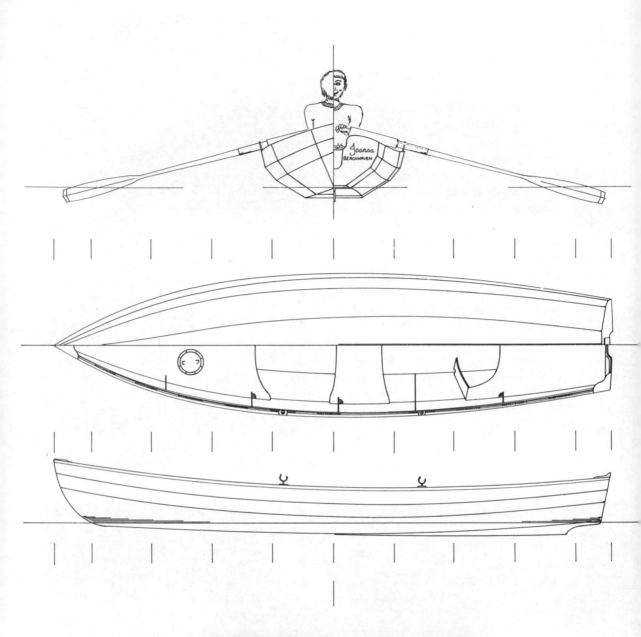

Joansa

LOA
4.70 m/15 ft 6 in

BEAM
1.18 m/3 ft 10 in

WEIGHT (approx.)
36 kg/80 lb

A Christmas Cruise with a Difference

Christmas in New Zealand is geared around the holiday, traditionally at the beach, perhaps in the bach, or for many like myself, away in the boat. Now, the boat typically ranges from a half-cabin runabout or a trailer yacht, through bigger and bigger craft, up to those top-heavy chrome-and-plastic 50-something foot 'gin palaces', vessels the like of which few people can ever hope to own, and which seem to get more beyond my pocket every time I see one — not that I would want one!

One Christmas, though, I couldn't even lay my hands on one of those for a bit of time away. We had boats but only small ones — very small by some standards. After some discussion, Jan and I decided that a cruise was a cruise, and we'd have a different type of maritime holiday.

Among our 'fleet' was *Joansa*, 4.7 m (15 ft 6 in) long, and more than capable of keeping pace with Jan's plywood touring kayak *PiWi*. (The name came about because of our misjudgement of the time necessary to cover a given stretch of water, a large cup of tea prior to setting off, and the fact that the then unnamed canoe lacked even the most basic of toilet facilities!)

Daughter Sarina, then aged seven, could not be expected to paddle her canoe too far at a time, and with Jan having only a few short trips in her own log, we were not keen to bite off too much in terms of distance.

We chose to tour Kawau Island and the 'Motu's', the chain of islands to the south, all a conveniently short distance from one another, and all within range of Sandspit where our car would be left in the secure carpark.

I left home a few days before Jan and Sarina with the rowing boat and my own kit on the roof-rack of the stationwagon, driving up to Sandspit with the idea of doing a little exploring before the family arrived. Unloading the boat and packing the gear took only a short time, the little 2 hp Honda outboard — my insurance for meeting time deadlines if the weather cracked up — stowed out of sight under the centre thwart (it spoils the macho image). With two twenty-litre jerrycans of fresh water, and all of my gear secured and well waterproofed in case of spray, I was off.

There was a steady north-east headwind blowing as I rowed out past the lines of moored boats and into Kawau Bay, my little cruiser rising to the swells and butting through the crests as I pulled steadily at the oars, making good progress towards Mansion House Bay on Kawau Island. I held over to the north of my course a little to stay out of the worst of the wind-generated waves for as long as possible, and was passed by several ferries, a couple of which went a bit out of their way and slowed down to have a look at the unusual sight of a lone rower out at sea.

I was timing all of my passages to get an idea of what distances we could cover, and in spite of the headwind the 4.2 nautical miles of this leg took only an hour and six minutes. I've taken a lot longer to sail this distance in a much larger boat!

Kawau was once the residence of New Zealand's first Governor General, and Sir George Grey's presence is still felt in the lovely old mansion, now restored and aptly known as Mansion House. The magnificent gardens contain many exotic trees and a unique mixture of native and foreign fauna; this is one of the very few places in New Zealand where kookaburras and wallabies mix with tuis and wekas.

When I arrived at the beach I joined the troops of tourists at the tearooms, watching in amusement a belligerent and overweight fellow who, in spite of being warned not to feed the birds by hand, had his finger well bitten by a weka. What was not so funny was this clown's immediate demands to 'see the boss here' and his enraged threats to sue! I felt that the superb Devonshire tea was not worth the aggro, and quietly slipped away.

Gliding through the anchored holiday season fleet and off down towards the head of Bon Accord Harbour, I entered another world, one where many people were interested enough to stop and chat to a lone rower in a long and graceful boat. Odd that — cruising in a big boat can be an exercise in being studiously ignored by all others, but cruising in small boats seems to arouse a more sociable brand of curiosity in many people. I've had more cups of tea than I can remember and some memorable dinners as a result of a conversation over the rail of an anchored cruiser.

Packing up Joansa *before setting off from Sandspit with the family. Seven-year-old Sarina is ready for adventure.*

Once upon a time there was a camping ground in Bon Accord, but with this gone it was either find an obviously unoccupied piece of ground or seek permission from a resident. With only myself to worry about this time, I decided to go exploring; I'd worry about the campsite later.

I rowed to the head of the harbour and across the tidal flats, mentally noting the creek that disappeared into the bush behind the sandbar at the end; and around into the other arm, noting as I went a couple of spots just nicely sited for a pup tent and camp stove. I rowed back through the anchored fleet once again, around into North Cove, and onwards along the shore. After rowing as far as the light at the northern end of the island, I quietly eased my way back along the coast, sneaking through the rocks and shallows, and stopping to chat with sunseekers and bach residents.

Late in the afternoon I cruised back up Bon Accord Harbour and across the mudflats at the head of the eastern arm, the sun setting over the transom as I rowed. With the tide being high in the morning and afternoon, I was able to row right up to the grassy bank below my chosen spot

and tie the boat to a tree, secure in the knowledge that the boat would be left dry all night — no need for an anchor watch here!

My camp was a very cosy one, to say the least! There was just enough flat ground for my tiny tent, with a comfortable seat in the crook of a tree's roots, and a grand view down the harbour, a view made all the better for the song of the tuis in the flax all along the shore. After a dinner made superb by a sharp appetite, it was very satisfying to drift off to sleep, so remote and peaceful, and yet only a couple of hundred yards from a harbour desperately crowded with seasonal cruisers in their much bigger craft.

Next morning was the sort of day just made for rowing — lightly overcast and mirror calm. Up and out early, my body clock adjusting easily to retiring and rising with the sun, I was fortunate enough to be invited aboard Julian Godwin's *ODTAA* (*One Damn Thing After Another* — a comment on some of the problems encountered while building her!) On discussing the search for campsites, Julian mentioned that he owned a patch of land around behind the sandbar at the end of the harbour, and that my family and I were welcome to use it as a base if it suited us.

Leaving the very sociable trio on board *ODTAA*, I headed off around the southern coast of Kawau to Bosanquet Bay, looking into South Cove, and stopping off on The Haystack for lunch and a swim on the way back.

Solo touring in a rowing boat is the nearest thing to being the proverbial 'fly on the wall' that one can be. With an almost silent means of propulsion, and a low and inconspicuous slip of a boat, people tend not to take much notice of you. Sunbathers are among the obvious surprisees, of course, but there is, for me, an odd feeling of privilege as I drift unnoticed through the lives of people on or near the water — a sort of quiet sharing that I don't find anywhere else.

It was back to Bon Accord Harbour after lunch, and a lazy afternoon tea in the shade of one of the Mansion House palms. Early evening saw me following the tide into the inlet and up to Julian's 'camp'. After an exploratory walk I set up the tent for the night. Dinner was a memorable one this time, as I was off home on the morrow to pick up the family. I indulged in a big spread to use up the perishable stores — battered onion rings, pepper steak, cauliflower au gratin and potato croquettes, with a glass of red wine, followed by plum pudding and custard. There is no problem cooking meals like these on a one-burner stove if you plan well, and good food makes the enforced simplicity of life in a tiny tent much more pleasant.

Back home the next day I picked up the crew, and 24 hours later we were on the way back to Sandspit, *PiWi* travelling inside *Joansa* up on the roof-rack, and the stationwagon well loaded with gear. By early afternoon we were all ready on the beach, the boats loaded, the motor in its operational position, our lifejackets on, and the car secure in the carpark. The tide was with us so we set off.

After an hour or so, Jan, being a real beginner at touring in a kayak, found the strong afternoon breeze a problem, so I picked up her already prepared towrope and fired up the little outboard to take us into the shelter of Bon Accord Harbour. (I am only a purist on fine days!)

Getting into Mansion House Bay through the tide race off the point was a little exciting. *Joansa*

has only about 250 mm of freeboard, and *PiWi* much less, so even small waves assume considerable importance. Not to worry though; we bounced over the chop in a very reassuring manner, with *PiWi* taking only a little water over the foredeck, and *Joansa* remaining completely dry in spite of her heavy load of camping gear and provisions.

We celebrated our arrival at the 'overseas' port with tea and scones at the tearooms, Sarina intrigued at the boldness of the wekas as they scavenged for scraps, and later enchanted with the shy wallabies in the distant shrubs. After our break it was a leisurely cruise up the harbour to our refuge behind the sandbar, where we set up camp for what was to be a three-day stay.

This was a wonderful spot. Sometimes we walked, exploring the roads and tracks along the south side of Bon Accord, making it right around to 'civilisation' at Mansion House on a couple of occasions, as well as skinny-dipping on the tiny beaches of Kawau's east coast. There are banana trees up in the valleys, as well as palms and macadamias, all planted by some keen botanical experimenter, which added a tropical air to the scene as we walked. At other times we took to the boats and investigated the coastline and its many historical relics.

On day three we took to the boats again, out past the point and across the six miles or so to a sandy stretch of beach near Takatu Point, hauling the boats up by the interesting-looking reefs that I had found on a previous voyage. The objective was to introduce Jan and Sarina to the joys of skin diving with basic flippers, mask and snorkel.

After a few minutes covering the principles of breathing, diving, clearing the snorkel and keeping the mask from flooding, it was little Sarina who stripped off and, kitted out with her lifejacket, kicked out along the rock shelf, surfacing with a squeal and a splutter to shout excitedly about the fish she had seen.

It wasn't long before Jan and I were into the water, and after some practice for Jan we set off to see what 'the kid' had found. What a wonderland! We had selected a near-perfect site for the first dive (more by accident than good management, but I was pleased to take the credit), with sheltering fingers of rock dropping vertically

Jan in PiWi, powering along nicely.

into about two metres of water, and a sandy bottom that gave us the experience of drifting weightlessly over a sun-dappled world populated by a wide range of marine life.

Prosaic to many, the little blue maomao, sprats, spotties and other reef fish were, to us, the magical inhabitants of a brand new world that hid beneath the surface of a sea that had, until then, seemed to my family just something that hubbie and dad liked to go sailing on. We had the incredible experience of hovering motionless on the surface as a group of small rays 'flew' past in formation below us — breathtaking! We've carried masks and flippers on many of our trips since, but for Jan and Sarina that first time will always be special.

Getting back across the channel was a battle, the wind against the tide kicking up a big sea, but Jan's rapidly improving fitness enabled her to push *PiWi* through it. It was not until we were resting in the calm water off Speedy Point that the wake of a thoughtlessly (and illegally) driven launch caught the canoe broadside on, and gave Jan only enough time to produce a few

startled (and unprintable) words before she was swimming.

Rescuing the casualty took only minutes. The stability of the touring kayak and the inexperience of the occupant had made self-righting impossible, so I hauled Jan over the gunwale of the 'mothership' and towed the swamped canoe to the shore a few yards away. We'd planned a shower anyway so it was off to the Kawau Island Yacht Club, a facility well worth the small subscription if you're cruising in the area, and washed ourselves and our several days' worth of dirty clothing in the showers before returning to the camp.

A move had been planned for the next day, so with near-perfect weather it was up at the crack of dawn, Sarina packing her own gear and tent while we got the rest of the camp stowed and the area tidied. With breakfast cooked and eaten, and the boats organised, we were off, heading south towards our next planned camp at the island of Motuora.

Cruising easily at a lazy three knots, I kept the heavier rowing boat just ahead of Jan to provide a smooth patch in order to conserve her energy. There was only a tiny breeze, so once we were out of the tide there was little wave action. After stopping at the distinctively shaped Beehive Island for a leg stretch and a cuppa, we stroked off towards the next in the chain of islands, easily overtaking several sailing boats on the way, enjoying the almost effortless progress in the calm seas.

Our next stop was Moturekareka Island. Exploring the bones of the old coaster *Rewa* that shelters 'Old Snow's' refuge was an interesting way to entertain ourselves while we took our break. The wreck towered impressively over our little craft, and provided a patch of perfect calm for us to brew a cup of tea to go with lunch.

From here to Motuora is another six miles, not far for many, but for an already tired beginner the island looks both remote and small from the eye level afforded by a very small boat. But Jan stuck at it, in spite of the wind getting up, and

'*Haul away, Daddy!*' A Sarina's-eye-view from the sternsheets as we arrive in Schoolhouse Bay on Kawau Island.

soon the distance was shortening noticeably as the cliffs at the northern end of Motuora loomed up impressively. We approached the island farm where we were to camp at the Department of Conservation's 'maritime motor camp' on the western side.

We booked in with the resident ranger, who came down to the beach to have a look at our transport; we were pleased with his expressions of admiration for our boats and the ground we'd covered thus far.

Motuora is a great place to stay. The campground is run by the ranger, and has very basic facilities in the way of toilets and a shower. The latter is sort of hot, late on a sunny afternoon; I say 'sort of' because the water is siphoned from a dam at the top of a hill above the camp, the black plastic hose being heated by the sun. However, bits of the pipe run through the cold shade under the many shrubs along the way, which makes showering something of an adventure!

As a farm park there are lots of big black cattle beasts about. Although I am a farm boy, for Sarina the experience of watching steers being unloaded from the supply barge, seeing the reluctant animals sliding down the steel ramp on their butts, and helping to ease them off the beach and into the appropriate paddock, was nearly as earth-shattering as watching the goats being milked in the evening, and realising that the milk she'd had on her cornies that morning had come from the very same nanny! It says a lot for the little girl's adaptability that next morning's breakfast went down just as readily, in spite of a muffled 'baaaaaa' from her dad.

We have special memories of this camp — sing-songs around the fire with other campers (just enough to be sociable, not enough to be crowded); the kids going off with Ranger Neil for the day to 'help' with the fencing; me teaching the same bunch of kids to slide down the hill using an old tarp as a sledge; bringing in the goats for evening milking; sunbathing nude at the beach over the hill; and lots more snorkelling.

One memory stands out though. Our tent was on a bank only a few feet from the high-tide line, facing west towards the cliffs at the end of the Mahurangi Peninsula, a few miles away across the water. I awoke before dawn one clear morn-

There are times when impending bad weather makes a fast passage desirable. Here we all are in Joansa, *assisted by Mr Honda pushing us along at about 5 knots on only quarter throttle. Out of the picture is Jan's* PiWi, *following along on its towline.*

ing and lay watching the full moon sinking slowly towards the cliffs. Jan woke to share the moment as the sky slowly turned grey, the stars winking out one by one as the palest of blues became evident, while the sun rose behind us, turning the cliffs a brilliant red as the moon sank behind them.

All too soon the time to leave came around. The wind had been in the south for a couple of days and was blowing pretty hard, keeping the boats up on the beach to avoid the very rough seas. But deadlines are deadlines, and employers a sorry intrusion into the real world, so it was time for the gear to be packed and for us to leave paradise.

By daybreak next morning we were off and away across the straits to be into sheltered waters before the wind really got on with the day's work. For this leg of the journey I had Jan and Sarina aboard *Joansa* with me, lifejackets on, and *PiWi* following behind like a friendly dog on her lead.

Little Mr Honda pushed us easily through the metre-plus chop that the stiff wind had kicked up, the boats riding well with only a little spray aboard. We made good progress, and it seemed only a little while before we rounded Mullet Point into the shelter of Scandretts Bay.

From there on it was a thoughtful and uneventful voyage along the coast to the car at Sandspit, and the drive back to Glenfield and home was a bit like getting into a cold pool an inch at a time, reality slowly seeping in around the edges.

We will be going back to Motuora, and hope to enjoy the island with young Brendan, our new family member, who is bound to provide us with a fresh point of view on a place that we have come to love.

Incidentally, if you want to visit this little bit of heaven, no longer a farm park but gradually being reforested in native trees, phone the ranger; there is a modest fee and you will need to book ahead.

Design Fourteen: *Penguin*

A four-berth trailer yacht with character

Several years ago, *Sea Spray* magazine ran a design competition, the criteria calling for a twenty foot trailerable yacht suited to family sailing and home building. I figured that there would be any number of entries featuring plywood boxes with three-quarter sloop rigs, so I thought I'd take advantage of the traditional appearance of my pet construction system to produce a nice little trailerable gaff sloop — *Penguin*, as she became — that would stand out from the crowd.

Well, she did that. I gather she caused some serious dissension among the four judges who, after two meetings, played safe and gave the nod to a high-performance 'go faster' that was a lot more 'mainstream' than *Penguin*. I did get an honourable mention, and she created a lot of interest when the drawings were published in a subsequent issue of the magazine. Possibly the best description of the design is the one I wrote in the submission to the contest judges; I've reproduced it here so you can judge for yourselves.

Penguin is based on its predecessor *Rogue*, enlarged and re-proportioned to suit the *Sea Spray* competition criteria, and I am very pleased with the resulting boat. Each of the criteria seems to me to have been well met, and the boat seems to be practical, promising good performance that would be achievable by a person of limited experience with a small budget.

She is a little longer than the specified length, but that has allowed two benefits: one, an improvement in the interior layout; and two, an improvement to the motor mounting. The additional costs of this extra 300 mm are not great, and the benefits are considerable.

I like to snuggle up with my wife after a day's cruising, and that luxury-size forward berth, complete with its own entrance (the forward hatch), comfortable sitting headroom where

needed, and personal locker space, has got to be a plus. It's almost as big as a queen-sized bed!

The quarter berths are long enough to accommodate an adult, while still allowing another to sit at the galley or table. They are wide enough for comfort, and high enough for even a very wide hip to turn over under the cockpit seats.

Penguin's cockpit self-drains through the motor-well and transom, ensuring the rapid return of unwanted briny to where it belongs — none of those bath-plug sized drains here! This, and the bridgedeck at the forward end of the footwell, should keep all but the most catastrophic swamping from disturbing the watch below.

The galley is small but benefits from the position of the table. It keeps the cook out of the traffic, has ample locker space, is safe (hot spills are unlikely to reach the cook), can be fitted with a pump supplied from a couple of twenty litre jerrycans, and is unobtrusive when not in use.

If she is to be a 'family' boat then she will need a dedicated loo. Given a choice, I'd fit a Portaloo under the bridgedeck and enjoy the bigger main cabin, but my wife tells me that I'd better think again. So, *Penguin* has a dedicated heads compartment, wide enough to reach behind while seated, high enough for sitting headroom, and with enough legroom so one can pull one's trou up afterwards. You'll notice that it even has room in the lockers for the pile of old *Readers Digests* that seem to infest the smallest room in most people's houses.

Despite being the main route from the 'saloon' to the 'forward stateroom', it does afford a lot more privacy and comfort than the plastic bucket I use in some of my less well-appointed boats.

The tabernacle, as drawn, is common on English yachts, and on a character boat such as this does not look out of place. Its big advantage is that the boat can be rigged by one person. The stability offered to the mast by the high tabernacle sides makes walking forward hand over hand under the mast, then holding it up with the jib halyard while the forestay is connected, a piece of cake. The 'gaffer', of course, is the easier of the two rigs in this respect, but neither should be a problem.

Question: When is an outboard not an outboard? Answer: When it is mounted inboard. With the transom sloped the 'proper' way, and the cockpit deep enough for real shelter and comfort, it would not be practical to hang the motor on an exposed bracket out where it would take an acrobat to reach it. On the other hand, the boat is too short to use almost half of the footwell for a motor well as we would normally see. This off-centre well, with the motor folding up through a slot in the transom, keeps the motor's powerhead and controls safely accessible, does not intrude unduly into people's space, is closer to the centreline than an outside bracket would be, and doesn't cost nearly as much as the fancy fold-up stainless steel outboard mountings.

A small family cruiser needs considerable attention to stowage. By the time the skipper, the mate, and a couple of kids have all brought their essentials aboard for a week away, most small boats are dreadfully cramped.

Starting up at the sharp end, *Penguin* has a large anchor well to take care of the main anchor. In a boat of this size I'd carry a 7 kg (15 lb) plough, ten metres of 9 mm chain, and 50 metres of 9 mm nylon warp, it will all fit easily in the well.

Down below, the forward bunk has a stowage net above the foot end, personal lockers each side of the head end, and considerable storage underneath. The heads has its own locker for essentials (and reading) while the main cabin has shelves under the side decks, stowage behind the backrests plus down the quarter berths, lockers under the berths and the galley bench, plus a big space under the bridgedeck. There is also a useful space under the cockpit sole if the builder wishes to make a shaped box with runners on to slide in like an oversized drawer.

Out in the cockpit there are large lockers for sails, warps, the spare anchor, wetties, fenders, and the boom tent, while there is a special place, draining overboard, for the motor's fuel tank and the stove's gas bottle. There are little 'caves' built into the cockpit coamings to take the small odds and ends of gear likely to be required while on passage.

In terms of safety and seaworthiness, the boat is solidly specified in hull and rig, has a very high righting moment, a self-draining cockpit of moderate volume, a sealed centrecase with minimum weight in the centreboard (I am not a fan of having a lot of ballast weight built into a board which can be raised, and which may swing back up uncontrolled in an extreme knockdown), a motor which should remain operable in extreme conditions, and a hull form which its predecessors have proven will carry sail in high winds and is capable of making headway to windward in pretty serious conditions.

My experience with similar previous designs has been very encouraging, and I am confident that this design, properly outfitted and in reasonably sensible hands, would be as safe and seaworthy as one could wish from a shoal-draft vessel of this size.

With regards to the philosophy behind the design, this boat is number sixteen in a development series of which ten have been built, some in considerable numbers, each contributing to the development of the next. *Rogue*, a successful 4.45 m cruising dinghy, was the basis on which *Penguin*'s hull form and construction was based.

Penguin is built upright, her bottom panel made up in one piece and set up on a simple jig which holds the panel in the correct fore and aft curve. Next the keelson, centrecase, stem, spine, bunk fronts and cockpit seat fronts are set up, with the frames and bulkheads fitted across, and stringers fitted fore and aft.

Planking is fitted in 2.5 m lengths, lap jointed like a clinker-built vessel, and butt joined at the ends to save time-consuming scarf joins.

When the builder starts the interior, it is largely a finishing-off job as much of the accommodation is already there in the form of bulkheads, fore and aft 'fronts' and spine members.

Many boats have been built using this system, and complete beginners are producing nicely built boats in a remarkably short time. Few builders find it necessary to phone me with problems, and the boats are surviving well in use, proving strong and durable as well as light, simple and relatively inexpensive to build.

Penguin's design was influenced by many factors other than the competition's criteria; styling was one. The lapped-side construction gives the boat a traditional air, so I added to this a gaff rig with the advantages of low cost, a shorter spar built of wood (not hard to build at home), and a low centre of effort. I feel that it would be a better rig for cruising for those who do not wish to carry extra downwind sails, and gives the boat a character that the conventional sloop would lack.

I've given her a high ballast ratio, a hull with high form stability, and have done my best to see that she will be both safe and comfortable while having performance that would make her fun to sail, as well as the looks which a family would be proud to point to and say, 'That's our yacht.'

Penguin

LOA
6.4 m/21 ft

BEAM
2.24 m/8 ft

WEIGHT (approx.)
970 kg/2138 lb

SAIL AREA
Bermudan rig 20.7 sq m/
223 sq ft
Gaff rig 21.8 sq m/235 sq ft

BALLAST
450 kg/992 lb

HEADROOM
1.45 m/4 ft 9 in

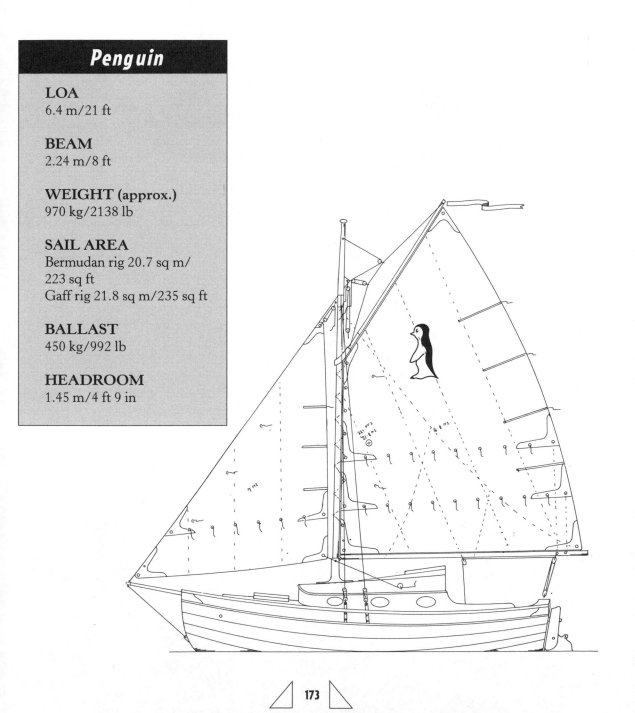

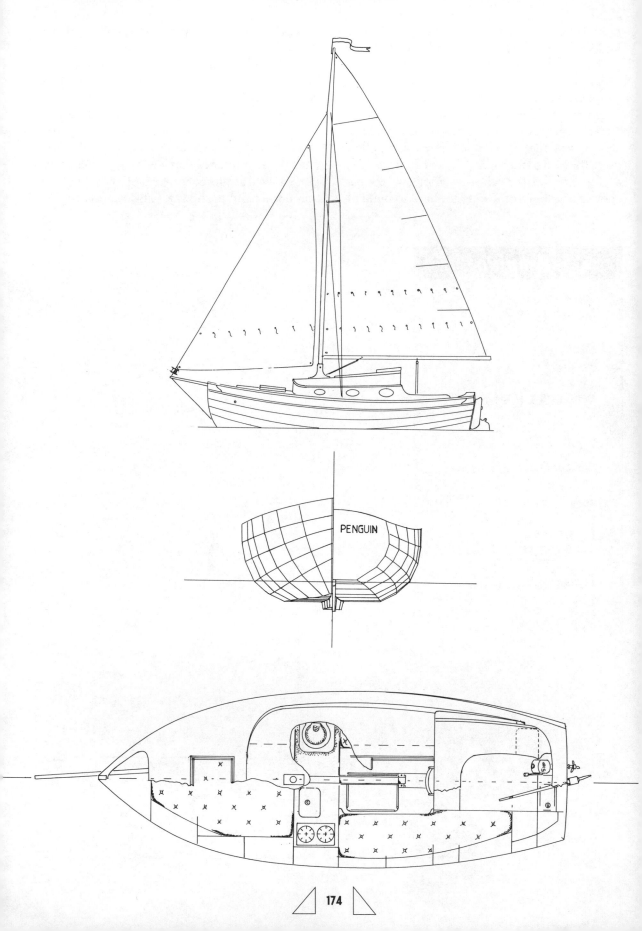

PENGUIN

Design Fifteen: *Sweet Pea*

A sports car with a pair of bunks in the back

In conversation with Tasmanian boatbuilder Stephen Dobson, I asked, 'What would be the next design you'd like me to add to my stock plans?' After some discussion he promised to give the matter some thought, and over some months of irregular dialogue the broad parameters for *Sweet Pea* evolved.

Stephen's home waters are the Tamar estuary from Launceston down past Robigana, where he and his family live, and out to Bass Strait on the northern side of Tasmania, an area of shallow water and strong winds, fast tides and steep choppy waves, broad reaches and secluded inlets — a pretty good imitation of paradise for a keen body with the right shoal-draught boat.

The idea was to produce a small cabin yacht a step on from the popular but ageing Hartley 16, a craft with a bit more speed, especially to windward in a chop, a self-draining cockpit, and a cabin that would seat four, and would not take water below if it came in over the coamings. The idea was to be able to knock about the bay with anything from one to five aboard, to race with two or three, and to cruise for a few days with two or — with the aid of a boom tent — four overnight.

She needed a lot of space 'outside'. A wee boat like this is primarily a daysailer, in spite of the cruising option. Race performance was a must. Stephen had thoughts of pumping out hulls from his boat shop for home completion, and the main market was envisaged to be those whose old Hartleys were getting too tired to be competitive, but who still wanted to have the excitement of a high-powered little boat, with the shelter of a cabin, and the light weight of a small trailer boat.

I sent a couple of drawings across the Tasman and comments came back by phone and letter. What you see here is the result. *Sweet Pea* is built from pre-shaped plywood panels taped and epoxied together (the stitch and tape way), and has almost no solid wood in her. She is simple in structure, very quick to build, light in weight, and extremely strong. There is little framing to take up inside space so accommodation has more space than you would expect. Her styling is slightly traditional, giving still more room, but the underwater lines are as 'fast' as possible, consistent with her daysailing and cruising roles.

With her large and powerful rig, I'd expect *Sweet Pea* to plane on most points of sail in a breeze, to be very good in light winds, and — with three rows of reef points — still be easily handled if caught out; with two or three reefs in the main, and the jib furled, I'd expect to be able to cope in really rotten weather.

Three metres of length in the cockpit means that the crew can be comfortable, get their weight in the right place, and trim her right (got to make the mark first!), or just stretch out full length and sunbathe. *Sweet Pea* is an honest attempt to produce a boat that will fill the needs of the whole family. Dad, crewed by Mum or a couple of teens, can race her; two, whether Mum and Dad, or the teens, can sneak off for a long weekend; and there is room for all four to sail across to 'that island' for a picnic.

As soon as my shed is finished, I'll be visiting Gordon at Plywood and Marine; it's a worry when you get hooked in by your own sales pitch!

Sweet Pea

LOA
5.30 m/17 ft 5in

BEAM
2.24 m/7 ft 4in

DRAFT
Board up 0.26 m/0ft 10in
Board down 1.35 m/4ft 5in

SAIL AREA
Excluding spinnaker
18.4 sq m/198 ft

WEIGHT (approx.)
Rigged, without stores
390 kg/860 lb

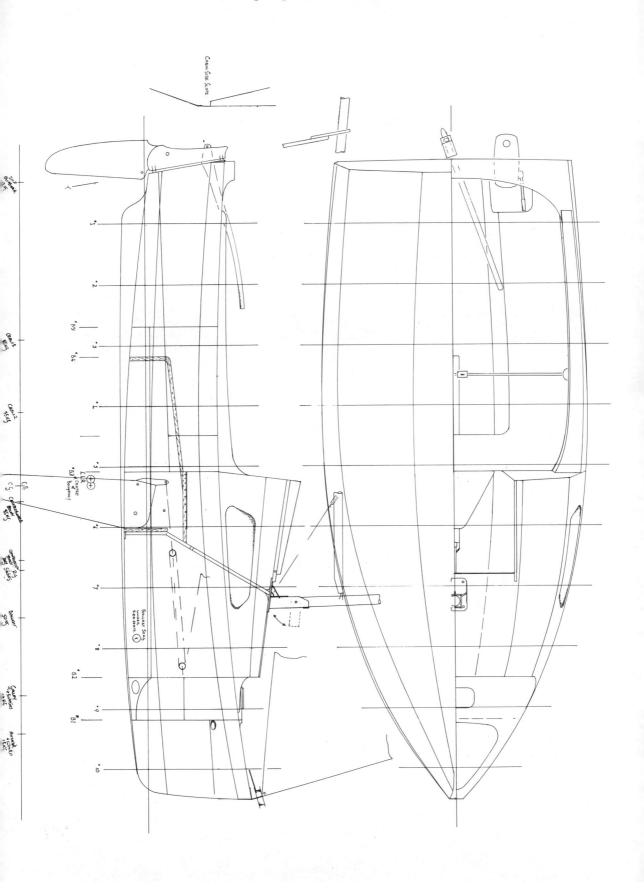

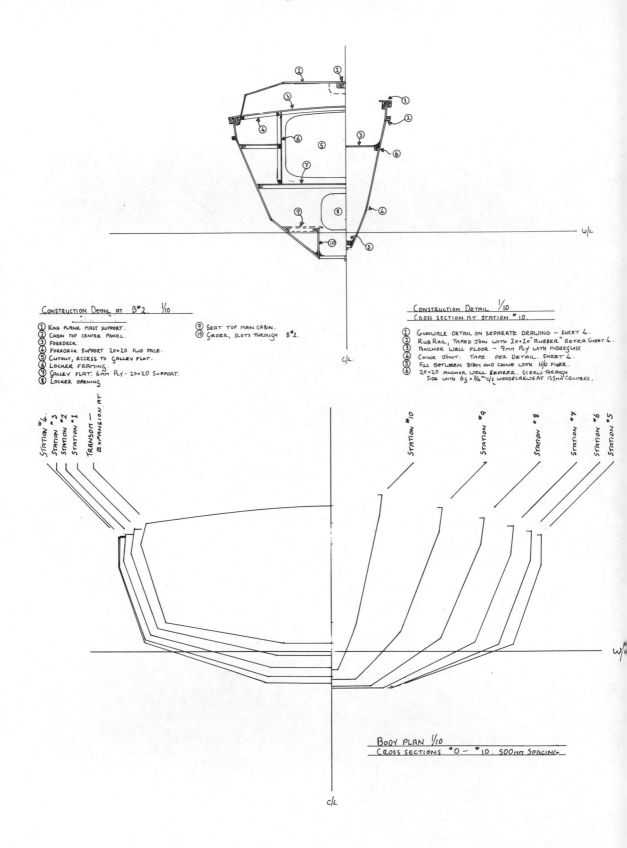

CONSTRUCTION DETAIL AT B#2. 1/10

① King plank mast support.
② Cabin top centre panel.
③ Foredeck.
④ Foredeck support 20×20 fwd face.
⑤ Cutout, access to galley flat.
⑥ Locker framing.
⑦ Galley flat. 6mm ply - 20×20 support.
⑧ Locker opening.
⑨ Seat top main cabin.
⑩ Girder, slots through B#2.

CONSTRUCTION DETAIL 1/10
CROSS SECTION AT STATION #10.

① Gunwale detail on separate drawing – sheet 4.
② Rub rail, taped join with 20×20 "rubber". Refer sheet 4.
③ Anchor well floor – 9mm ply with fibreglass.
④ Chine joint. Tape per detail. Sheet 4.
⑤ Fill between stem and chine with H/O filler.
⑥ 20×20 anchor well bearer. Screw through side with 6g × 3/4" c/s woodscrews at 125mm centres.

BODY PLAN 1/10
CROSS SECTIONS #0 – #10. 500mm SPACING.

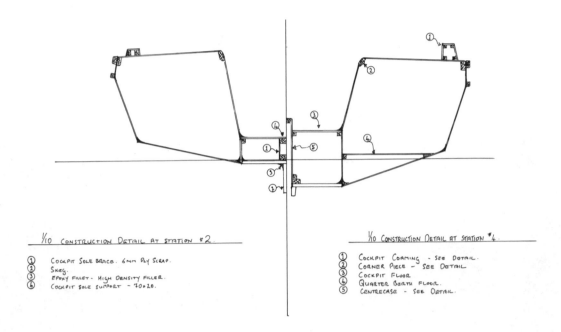

1/10 CONSTRUCTION DETAIL AT STATION #2.

① COCKPIT SOLE BRACE. 6mm PLY SCRAP.
② SKEG.
③ EPOXY FILLET - HIGH DENSITY FILLER.
④ COCKPIT SOLE SUPPORT - 70×20.

1/10 CONSTRUCTION DETAIL AT STATION #4.

① COCKPIT COAMING - SEE DETAIL.
② CORNER PIECE - SEE DETAIL
③ COCKPIT FLOOR
④ QUARTER BERTH FLOOR.
⑤ CENTRECASE - SEE DETAIL.

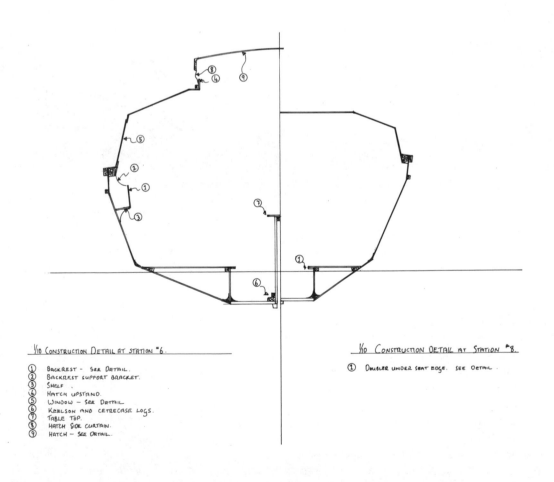

1/10 CONSTRUCTION DETAIL AT STATION #6.

① BACKREST - SEE DETAIL.
② BACKREST SUPPORT BRACKET.
③ SHELF .
④ HATCH UPSTAND.
⑤ WINDOW - SEE DETAIL
⑥ KEELSON AND CENTRECASE LOGS.
⑦ TABLE TOP.
⑧ HATCH SIDE CURTAIN.
⑨ HATCH - SEE DETAIL.

1/10 CONSTRUCTION DETAIL AT STATION #8.

① DOUBLER UNDER SEAT EDGE. SEE DETAIL.

Design Sixteen: *Rifleman*

A runabout in the classical style

All of the frames for a Rifleman, *cut out and ready to fit the reinforcing pieces.*

The lapstrake sides and classical looks of *Rifleman* were more or less an accident. Frank Perry had been on the beach at a popular resort when he had seen Roger Davies in his beautifully presented *Rogue* sailing dinghy.

Intrigued by the boat, Frank was soon talking to the owner, and boatbuilder Roger, keen to make a sale, soon had him out on the water. Frank bought a set of *Rogue* plans from me, and intended to have his new friend build the boat the following winter. However, after some study, he decided that a craft intended primarily for sail and oar would not be fast enough to get him to and from his favourite fishing spots along Wellington's iron-bound windy coast, so after

several toings and froings on the phone, we decided to design a completely new boat to suit.

My client described himself as 'not as young as I once was', and arthritis meant he couldn't tolerate a cold cramped seating position for long. His boat needed to be faster than *Rogue*, which could be pushed with an outboard but whose shape was wrong for planing speeds with a bigger outboard motor. The advantages of the unusually light weight and the consequent ease of handling were to be retained, along with the practical interior, ease of building, and — not least — the lapstrake or 'clinker' sides.

So, starting with a blank sheet of paper, I began drawing up *Rifleman*, an outboard motor-powered planing hull with a designed serviced speed under normal conditions of 18–20 knots. This should mean a top speed in flat water of 25-30 knots.

To obtain this performance with the modified dory hull form I used for *Rifleman*, a 20 hp motor should be adequate for those cruising two-handed, while a 25 hp should be quick enough for most, even when loaded up a bit. At the gentler end of the scale an 8 hp motor should push her along nicely, and a 15 hp would plane her comfortably with two aboard. One has been fitted with a 4-stroke Honda 15, and is a particularly smooth and quiet boat in use — and it doesn't leave an oily rainbow sheen on the water!

An oversize planing shoe enables the boat to carry a large load at planing speed on moderate power, or to plane under full control at comparatively low speeds. This gives low fuel consumption and easy handling, even in quite rough conditions.

The steeply rising bottom panels soften the ride and give good directional stability, while the long fine entry eases through the waves, and keeps the spray down in a way that the more common wide beam can rarely achieve.

Inside there is room for four to ride in comfort, and to fish without tangling lines. There is adequate storage under the foredeck and under the centreline of the seats, while the space under the side seats and forward thwart is sealed off, providing sufficient air tank buoyancy to float her stable and baleable when fully swamped. This space can be accessed through plastic hatches if required, and is a good place to keep bait, spare clothing, matches, food and other essential small items. (Don't put the bait in the same one as the food!)

Steered from the console on the centre thwart, the boat is easier to trim, and will generally provide a more comfortable ride with the weight out of the ends. With the motor's normal remote control kit, and a simple cord and pulley steering system, it is possible to rig the controls without having to spend too much on hardware.

Frank intended to launch the boat from beaches and rocky shelves that were barred to

Eric Ward, then in his eighties, built this Rifleman *for use on the Waikato River. With its 15 hp 4-stroke outboard motor, it provides a smooth and super-quiet ride for Eric and his friend.*

him with his heavy fibreglass speedboat, and looked forward not only to the convenience of the smaller boat but also the economies of the 18 hp Tohatsu outboard motor he'd chosen, not to mention the prospect of owning a more economical towing vehicle.

Towards the end of the design process, a couple of my letters went unanswered, and when a telephone message got no response I carried on with the drawing process but did not send the plans off for fear of them going astray. (Frank had already paid for them so I figured that, one way or another, he'd get back to me.)

Two or so years later, and after selling quite a number of plans, I heard from a very apologetic Frank. Roger had talked him into a more conventional 4.3 m (14 ft) deep vee from a very well-known designer. They'd built it, and it turned out that the thing had been less than adequate in the conditions that prevailed in Cook Strait! There was nothing really wrong with the boat — it was just the wrong one for the job!

With several *Riflemen* in the water by then, I was able to reassure Frank that the boat performed pretty much as advertised, and sent off the plan set. He's building this one at home by himself, and I hope to hear from him sometime to see how she has performed.

Rifleman

LOA
4.5 m/14 ft 9 in

BEAM
1.5 m/4 ft 11 in

WEIGHT (approx. bare boat)
90 kg/198 lb

SPEED RANGE
Up to 30 knots

POWER
7.5–30 hp outboard motor

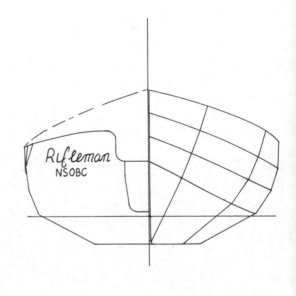

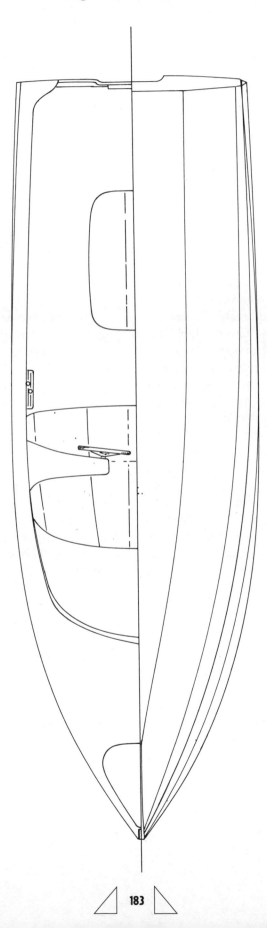

Design Seventeen: *Fishmaster*

A specialist fishing dory

I was sitting one day with editor Bill Kirk in the office of *New Zealand Fisherman*, chatting about the hypothetical 'perfect fishing boat'. Much to my surprise, neither Bill nor his deputy, John Eichelsheim, rated speed very highly. Of concern was the ability to maintain an economical cruising speed of about 25 knots in a variety of conditions; she should be stable at rest, unlike many deep vees; have a cuddy cabin big enough to seat four with reasonable comfort; perhaps a small cooker; and space for a portable loo.

The cockpit should be large with no steps in the floor, storage for rods under the gunwales, and bait and gear away in lockers, and both jokingly agreed that there should be at least one corner for each fisherman to brace himself against. This lead to a discussion as to hull form, which lead to the conclusion that a boat of octagonal shape with a cabin in the middle would seem to have some advantages!

I toddled off home and drew a proposal for a 5.5 m (18 ft) craft based on the successful *Rifleman* light runabout. I've used this shape several times, and find them a good compromise between the soft-riding but power-hungry deep vees and the elongated flattie normally branded a 'dory' by people who should know better. The same narrow, flat bottom and medium deadrise angle have been used, but a straight side and a more conventional profile have been given to the topsides. The trick with these boats seems to be to keep the entry very fine at the waterline, with enough flare just above to give her the reserve buoyancy to stop her from burying her nose in the back of a wave and broaching when running downwind.

Fishmaster looks pretty much like most fishing runabouts, but will carry a lot more weight with a given amount of power. She will ride smoothly if trimmed flat, and has lots of storage for gear. Intended for 35–70 hp motors, speed will not be a problem with anything bigger than a 50 hp motor, even with a big load; even the 35 hp motor will get her along pretty well in most circumstances.

Building her will be no big deal. One guy came to me on a Friday with a story about a pile of 9 mm ply, an old 50 hp motor, and a need to get six people plus dive kits out to some interesting wrecks. He was back on Monday with a cheque for the very basic set of plans I'd drawn for a longer, skinnier version of this, and three weeks later I heard through the grapevine that he'd just thrashed his mate's big fibreglass boat in an impromptu race around the northern harbour where they lived. It made me wonder whether the glue was properly dry, let alone the paint!

But still, there is nothing complicated here. It's no different to the other small boats in this book. A bit of care in fitting the self-draining cockpit floor, some patience in forming the panels around the sharp bow area, and some attention to detail in the cabin and cockpit trim should see the sort of boat that would have cost heaps if you'd paid someone else to build her.

Fishmaster

LOA
5.50 m/18 ft 0 in

BEAM
2.2 m/7 ft 3 in

WEIGHT (approx.)
Dry, with motor
320 kg/705 lb

SPEED RANGE
Up to 35 knots

POWER
35–70 hp (45 hp ideal)

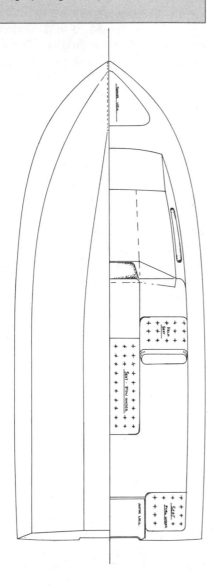

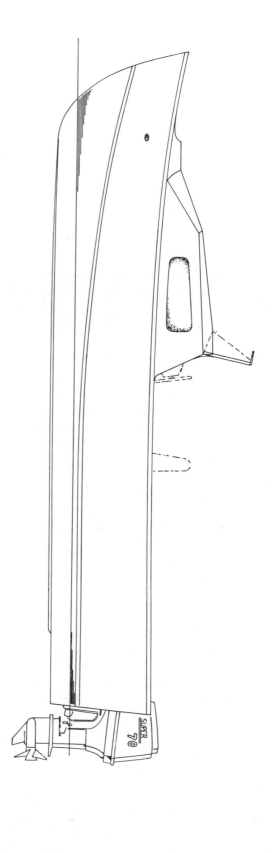

Design Eighteen: *Arwen*

An ocean-going cruiser for the realistic home builder

Like the little tender *Roof Rack* at the beginning of the design part of the book, *Arwen* (the Fairy Princess from Tolkien's *Lord of the Rings*) was designed for Marcus Raimon. Being a lad with somewhat itchy feet and an ambition to scratch them on a beach far away, he wanted a solid little cruiser that would take him in reasonable comfort wherever the wind blew. When looking at the tiny ferrocement yacht that had carried him up and down the coast, I felt that his idea of 'reasonable comfort' was a bit more basic than most, so I planned to give him an interior that would cater for an improvement in his standards. Besides, he's a handsome young fellow, and I would be very surprised if he didn't come home with a crew, even if he managed to leave without one! I wouldn't like to inflict the interior of *Roc* (Marcus's existing boat — a good name for one built of ferro!) on some nice lass.

So, with my client having loads of enthusiasm, beginner's building skills, and only a very thin budget, I had to be realistic in what I drew. Marcus's endless enthusiasm and skills as a scavenger could make up for a lot of shortcomings in other fields, but the structure had to be achievable.

To travel long distances in a small yacht, one has to be able to carry ample food, water, fuel, spares, a dinghy, anchor tackle, awnings, extra sails, and so on — lots of weight and lots of bulk! The amount of weight to be carried is a major factor in selecting the minimum size range of a long-range yacht, and allowing for an extra person meant that the miniature originally envisaged was not quite the ideal.

Looking at the areas in which our intrepid adventurer wished to indulge himself set some of the other parameters. Shallow draft would be an advantage. Strong windward ability, self-steering, a comfortable motion, and an interior that would be easily ventilated in tropical heat were considerations, as was the desire for traditional appearance. Her construction had to be as cheap as possible, and consistent with safety requirements and the resources available.

A skilled and determined amateur can produce a very good boat for very little in the way of money. It comes down to the value the builder puts on the labour involved. I would happily spend $100 on a fitting that would take me two days to make; Marcus probably wouldn't have the money, so would have to make a pattern and trot off down to the local foundry to have it cast.

Gaff rigged, with a pole mast from a local plantation of larch or cypress, galvanised 7 x 7 plough line for standing rigging, forged galvanised steel bottlescrews, and fittings cast from bronze at a local foundry using home-made patterns would make her much cheaper and not a lot slower than her stainless steel and alloy equipped sister. Down below, two second-hand Primus stoves in a hand-built gimballed frame would save more hundreds, while a little air-cooled diesel with a centrifugal clutch could be installed at a fraction of the cost of a 'real' marine diesel with its complex gearbox, cooling system and electrics (not to mention the big chunk of tax the government want from it).

With these economies, and a structure achievable by a comparative beginner, I drew the little cutter shown here, and although I have had an eye on the budget throughout, there is no reason for her to be anything other than seaworthy, fast, and very pretty.

Lin and Larry Pardy have made the small pilot cutter types, designed by (among others) Lyle Hess, very popular today, as they are well suited to short-handed cruising. I have used these capable little boats, with their long keels and beamy stable hulls noted for seaworthiness

Arwen

LOA (on deck)
7.75 m/25 ft 5 in

BEAM
2.90 m/9 ft 6 in

DRAFT
1.40 m/4 ft 7in

DISPLACEMENT (light)
3560 kg/7848 lb

SAIL AREA (working)
32.5 sq m/350 sq ft

and comfort, as a guide when designing young Marcus's boat. At 7.6 m long (25 ft) on deck, 2.9 m (9 ft 6 in) beam and 3.5 tons dry, *Arwen* has the room and carrying ability to provide for her occupants in reasonable style. There is full standing headroom in the main cabin; even cabinetmaker Radley Clark, about 1.88 m tall (6 ft 2 in) and prodding me to finish the Mk II drawings so he can build his own version of *Arwen*, will be able to stand upright under the closed hatch.

In a hull of this volume, a wide variety of interior layouts are possible. For those cruising weekends with a family on board, a cabin plan with a fair degree of privacy might be preferred, while for Marcus we have an interior much more flexible in terms of storage and accommodation to suit his plans for very long-range cruising.

Up on deck there is a cockpit long enough to lie down in, and the footwell is quite small in volume and drains straight out through the transom under the large locker at the after end, where a gas bottle, fuels, warps, fenders, and a third anchor are stored.

Low coamings keep the little water that may make its way up onto the side decks away from one's rear, and also provide mounts for the two bottom-handle winches, while access to the normally unused space under the cockpit floor is through a flush hatch in the cockpit sole — a good place to keep items such as awnings and shopping carts which are not required at sea.

Wide decks make moving about the boat both easier and safer, and the low bulwarks help to keep feet on deck where they belong. There is a well deck up forward, big enough to stow the main and bower anchor with warps and chain, making a navel pipe and chain locker less of a necessity. That's one less hole in the boat in an area which can be very wet!

I've drawn a double headsail sloop rig, these days commonly referred to as a 'cutter', and occasionally (and inelegantly) known as a 'slutter'. I have a strong liking for a gaff rig in a boat like this, and two of the three now under construction will be 'gaffers'. For the impecunious this is a good way to go, as the spars can be carved from readily available timber, and fittings welded from mild steel at local high school classes, then hot galvanised.

With the mast in a tabernacle the rig can be dropped by the crew; a boatyard's expensive crane is not required, enabling the boat to be taken into those quiet backwaters up creeks and under bridges where no one else goes — ideal for long stays for work or maintenance.

Built from treated softwood (white pine in New Zealand), using the plywood permanent bulkheads as 'moulds' with stringers over, and two skins of plywood moulded, she would not have the boxy look usually associated with plywood. Although this is not the very cheapest method of building, it means that all of the materials are readily available, and the skills necessary to build a boat to a good standard are within the capabilities of most people.

If I could write a testimonial for her, it would be: 'Tucked away in a quiet tidal backwater, birds singing around her every morning as the young couple who built her prepare to go off to work in the nearby town, *Arwen* dreams of the ocean passage to come as soon as the cyclone season is over. A voyage that will see them relaxing in the sapphire-blue waters of the Isle of Pines, or exploring the many islands and reefs of the Vava'u group, north of Tonga. She is a comfortable home for her owners, whether at anchor or charging along in the trade winds, and is a real reward for the effort put in to build and outfit her.'

Design Nineteen: Lightweight Launch

A sweet and simple coastal cruiser

My friend Colin Frankham was burdened by a large planing launch with a thirsty V8 motor that was not going to be appropriate to his budget when he retired a few years hence. He'd come across the Yanmar diesel outboard motors at the Boat Show, and was intrigued with the possibilities that they offered. I'd been pondering the same, and had in fact drawn a simple motorsailing freighter with the D27 version for a family in the Philippines. The low cost, very simple motor installation, extra room in the middle of the boat, and the confining of the motor, with its attendant noise and smell, to a small enclosure at the after end of the boat, were all obvious benefits.

Colin has a home on the edge of the water and has a lovely view into the sunset, but his jetty is high and dry at low tide, a consideration when drawing a boat to suit. Other criteria were a stack of half-inch plywood that he wished to use, a need for a cruising speed of nine knots, and a fairly traditional appearance.

Jeanette Frankham had been, in her youth, a demon tennis player; unfortunately, this had left her with knee injuries that made most boats difficult for her to live with. She needs a steadier motion than most deep vees provide at anchor, and boarding is not easy, so providing for these needs was high on the list. Colin has built more boats than some professionals, so the skill level of the client was not a problem. However, we did not want to produce something that would take an age to build, so the intention was to design a sweet and simple boat that would suit his style, without breaking his budget in terms of both labour and money.

We'd been amicably debating the possibilities for quite some time, Colin not being ready to build, the big 'fizzboat' not having been sold, and the need to make a decision being a long way off yet, when I felt the need to put on paper the images that had been forming in my mind. Those images had been getting in the way — there seems to be room for only one design at a time in there. So I set to and drew what became known as the Lightweight Launch.

With the Yanmar diesel outboard swinging its fine-pitched prop down aft she'd be uncommonly free of noise and vibration, and with no motor box to clutter up the middle of the boat there would be a feeling of real space in the main cabin. Although the motor itself is not cheap, there are considerable savings: the conventional shaft and prop, fuel tanks, exhaust, and engine beds are all unneeded, as the outboard simply bolts on to a sturdy transom.

Note that the slight step down into the main cabin from the self-draining cockpit is the only change in floor level, making moving about the boat very easy. Raised up to get the eye high enough to see clearly over the bow, the helm position is a comfortable one just forward of the sizeable galley. On the other side of the cabin, also raised to take advantage of the view from the large windows, is an L-shaped convertible dinette with storage underneath, and bookshelves under the side decks.

In under the raised foredeck are a pair of vee berths with a Portaloo™ or similar in between (although there is certainly room in the boat for a dedicated head if preferred). Also between the berths are steps that enable the anchor to be handled from the forward hatch via a self-stowing system and anchor winch.

In appearance she is similar to those fine, comfortable and efficient launches that were built throughout New Zealand during the 1950s — boats that are still around, and which even today form the mainstay of the serious cruising fleet here.

The Lightweight Launch is intended to do much the same job but on less power, with a lower fuel consumption and a greatly simplified construction. The utilitarian interior is but one aspect of the boat that reflects the uncomplicated nature of this little coastal cruiser, and anyone building her would do well to avoid unnecessary complications that would take the eye away from what should be a simple and graceful vessel.

As I write, Colin has the 'big banger' at the brokers, and we are still trading ideas. I guess there won't be a decision on his new boat until he is ready to start on that pile of plywood, but speculating on what might be next and how to do it is one of the real pleasures in boating design, so it's not really time lost. If Colin decides to do something different I won't be out of pocket; I've got a design that I like very much, which may never have been drawn otherwise, and anyway, the salty old so-and-so has taught me a great deal in the process!

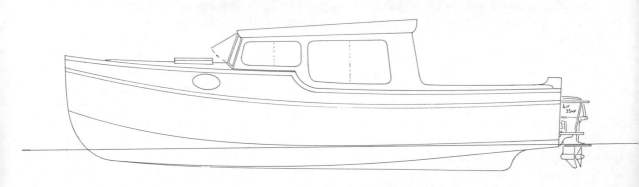

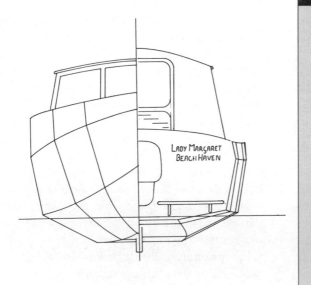

Lightweight Launch

LOA
8.6 m/28 ft 3 in

BEAM
2.8 m/9 ft 2 in

DISPLACEMENT (dry)
1600 kg/3500 lb

SPEED RANGE
Up to 11 knots

POWER
4 stroke petrol outboard, 35–45 hp
2 stroke petrol outboard, 40–50 hp
4 stroke diesel outboard, 27–40 hp

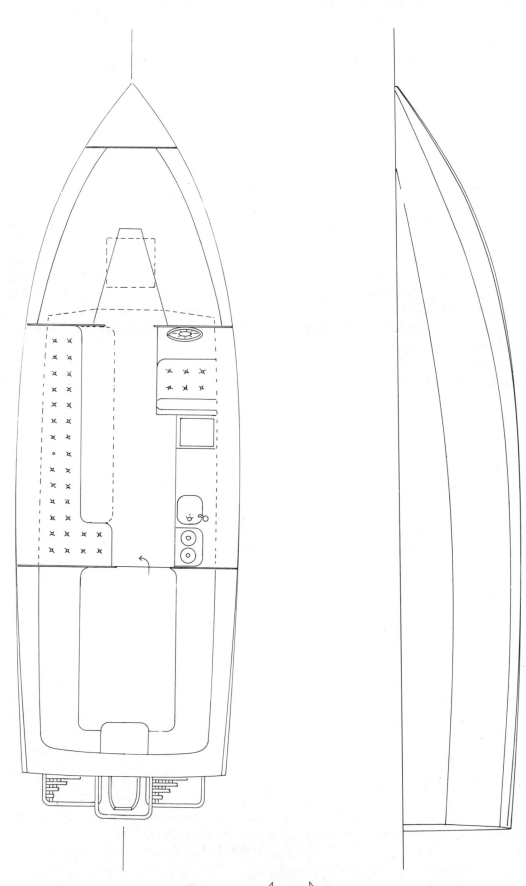

Design Twenty: *Queen Mab*

The littlest long-range launch

As with many of my designs, this one started with a phone call. My caller wanted a very comfortable motor launch in which he and his wife could spend the summers of their retirement. He had spent much of his working life as a maintenance engineer on the waterfront, and had long been in love with the stocky little ferries that bustle around Auckland's Waitemata Harbour, servicing the smaller routes.

He needed a craft that would be within his limited budget, within his skills as a boatbuilder (he described himself as a capable handyman, but I suspect he was being modest), within the length of his 9 m long (30 ft) workshop, and within his physical limits.

Long range was important, as was standing headroom in all major areas of accommodation. Speed was not a consideration, but she'd have to be able to cope with whatever weather came her way, as she could not be expected to run for it. A comfortable heads with a shower, and a well-equipped galley were also high on the list.

Accommodation was primarily to be for two. The double bed could be in the main saloon but had to be able to be left made up during the day without making the cabin unusable. There were also to be bunks for two grandchildren, but in a separate area so that Nan and Granddad could still have some peace of an evening.

Styling was to be reminiscent of those little ferries that had for so long drawn my caller's eye as they shuttled back and forth past his office window. For years they were known as the 'Blue Boats', and this was to be the base colour of the outside of his boat.

After having taken several phone calls to establish these criteria, there was a very shy addition to the list: 'Would it be possible for a boat like this to make the trip up to Fiji or Tonga?' Talk about a sting in the tail! I shuffled off to think about it.

Those who design small fishing boats know the problems of producing small vessels capable of carrying large cargoes at moderate speeds in all conditions, so it was to these, rather than the normal recreational launches, that I turned for inspiration. They use smallish engines with large reduction gearboxes, huge propellers, and very heavy displacement-to-length ratios. After rereading Robert P. Beebe's book *Voyaging Under Power* for the umpteenth time, and checking my conclusions in Dave Gerr's 'Propeller Handbook', I was able to tell my client (who by now had paid a small deposit to cover the initial work, so had changed in status from an 'enquiry' to a 'client' — a small distinction, but to one who needs to eat, an important one) that yes, it was a practical proposition — and *Queen Mab* was born.

At 7.9 m (26 ft) waterline length, around 5600 kg displacement, and an appropriate hull shape, a cruising speed of 6.2 knots would only require about a 14 hp engine. This figure can easily be converted into fuel consumption and back into miles per gallon, in this case 12–14 mpg using one of the more efficient modern diesels.

On the subject of diesels, the fuel efficiency of small diesels has improved considerably over the last few years. One manufacturer used to quote a popular-sized motor as using 230 gm of fuel per hp hour, and recently released a new motor of similar size with a claimed fuel usage of only 172 gm per hp hour. The difference would be about 30 percent, or a decrease in *Queen Mab*'s consumption at cruising speed from 78 litres per 24 hours to 55 litres per 24 hours! To achieve the required range at the better consumption rate would only need about 120 gallons (540 litres) — not hard to find space for in such a capacious hull, even when a generous safety margin is allowed.

After the usual round of very amicable phone

calls and proposal drawings, we had the little 'Pocket Hercules' that you see here. So I will take you for a tour …

In the fairly basic forward cabin there are a pair of 1.8 metre long berths that will sleep the grandchildren when they come to stay for the occasional overnight. There is standing headroom under the forward hatch which also provides ventilation. Additional airflow and a comfortable level of light come through the opening portholes in the cabin trunk — quite a cosy spot to relax in the afternoon while Granddad naps and Nan reads her book in the main cabin.

As we go up the companionway and into the wheelhouse, we note that the helm position has a very comfortable armchair, elevated so the person on watch has a good all-round view. There is a comprehensive tool and spares kit under the small area of raised floor, while the wheel is easily handled from a standing position amidships in bad weather, or — when the mooring needs to be picked up — through the starboard side sliding door.

When on passage it is an advantage to have a really comfy spot for everyone, and the short sofa on the starboard side is one of several such spots in this boat. The off-watch person could be very happy here as the miles tick steadily by, the door open to provide fresh air, and the overhanging wheelhouse roof keeping the hot noon sun out.

Under the wheelhouse sole a three-cylinder Lister, its already quiet note further muffled by extensive soundproofing and an oversize silencer, ticks over at a leisurely 1500 rpm, its 160 gallons of fuel in tanks either side of the motor under the wheelhouse floor.

Access to Mr Lister is by a removable panel in the wheelhouse, and further by the removal of the companionway steps at the fore and aft ends of the raised floor. This gives completely clear access, not only to the motor, but to the transmission, fuel, and cooling and exhaust systems, as well as the battery bank.

Stepping down into the main cabin past the dry exhaust trunking (heat exchanger built-in to provide hot water for 'domestic' use), we pass the heads on the starboard side; there is lots of room in here for the loo, handbasin and shower.

Opposite is the galley, with the stove sited so the cook doesn't wear the potful of boiling whatever if the vessel rolls. There is counter space to spare here, a good spot for someone who enjoys cooking in a spacious airy kitchen, and there's space for someone to dry the dishes while the other washes.

With only 8.5 m (28 ft) to play with, we opted for a saloon that seats four in comfort, rather than cramming too many into a small space. Conversion of the settees into berths is achieved by extending the foot under the bench seats out in the cockpit — a good place to stow the bedding when not required. The English call foot spaces like these 'trotter boxes' — a lovely description.

There is lots of window space in the saloon. The big windows, all in 12 mm Lexan™ and securely braced from inside, help make the small cabin feel much more roomy, and allow the occupants to enjoy a view almost as good as that from the wheelhouse.

To starboard, the settee converts into a double berth 2 metres long by 1.2 metres wide (the queen-sized berth comes in the 11 m version). This can be left made up without obstructing access to the after cockpit from forward. Although there are narrow side decks outside, it is not intended that they be used at sea.

Outside aft, there is room enough to handle mooring lines, or to board through the transom gate from the outside boarding platform. One could fish in comfort, or just sit and watch the wake — another pleasant spot.

Up on the cabin top there is a good stable 2.6 m (8ft 6 in) dinghy, complete with sailing rig. Hoisted up there on the derrick that swings off the mast which supports the tall exhaust stack, this little boat is well out of harm's way at sea, but can be launched and retrieved with ease. Also on the mast is a sprit-rigged steadying sail which, in combination with the roller-furled jib, gives the boat an alternative means of propulsion that prevents her from being completely helpless if her normal power source is out of action.

Climbing back down and forward onto the deck alongside the wheelhouse, we note high bulwarks with wide side decks inside, decks that run forward from one side of the wheelhouse to a workable foredeck and down the other side, giving a very secure area in which to work when mooring or tying up. With the forward cabin trunk forming a little raised 'island' in the middle, just at the right height to sit on, it is yet another comfortable place from which to sit and watch the world go by.

While the intended cruising speed of 6.2 knots is not fast by powerboat standards, any cruising yachtsman would tell you that the ability to cover 150 miles a day in a wide range of weather conditions is very quick indeed by sailing standards, particularly in a small and lightly crewed vessel.

Lack of length is this little explorer's only real vice; a beam-to-length ratio such as this will make her uncomfortably bouncy in a short beam sea. She will get to her destination okay in conditions that will frighten most, but I suspect that the owners will only want to do the trip once or twice before retiring her to the extended coastal cruising that she was designed for.

For the more adventurous, we are drawing an 11 m (36 ft) version that will have a noticeably more comfortable motion in bad weather, a sumptuous main cabin, the 'loo' forward of the wheelhouse, and a 3500 mile range. (And don't forget the queen-sized bed!)

Queen Mab is intended to be built from standard timberyard materials. Treated softwood and construction plywood are used throughout, making obtaining the materials very easy, and being a bit more environmentally friendly than using the fast-disappearing tropical hardwoods.

There would be nothing difficult in building her. With frames sawn from standard timber merchant sizes, a double skin of ply with a glass skin over stringers in a simple version of 'cold moulding', using the builder's choice of epoxy or phenolic resin glues, ordinary handyman's woodworking tools, a few galvanised mild steel fittings from a local engineering firm, and a lot of determination, this little ship is very achievable for many people.

Queen Mab

LOA
8.7 m/28 ft 6 in

BEAM
3.60 m/11 ft 10 in

DRAFT
1.12 m/3 ft 8 in

DISPLACEMENT (wet)
5600 kg/12346 lb

POWER
30-40 hp diesel motor, requires 25 hp continuous rating

ECONOMICAL CRUISE SPEED
6.2 knots

MAXIMUM SPEED
8 knots

THEORETICAL RANGE
At 6.2 knots, 2100 nautical miles

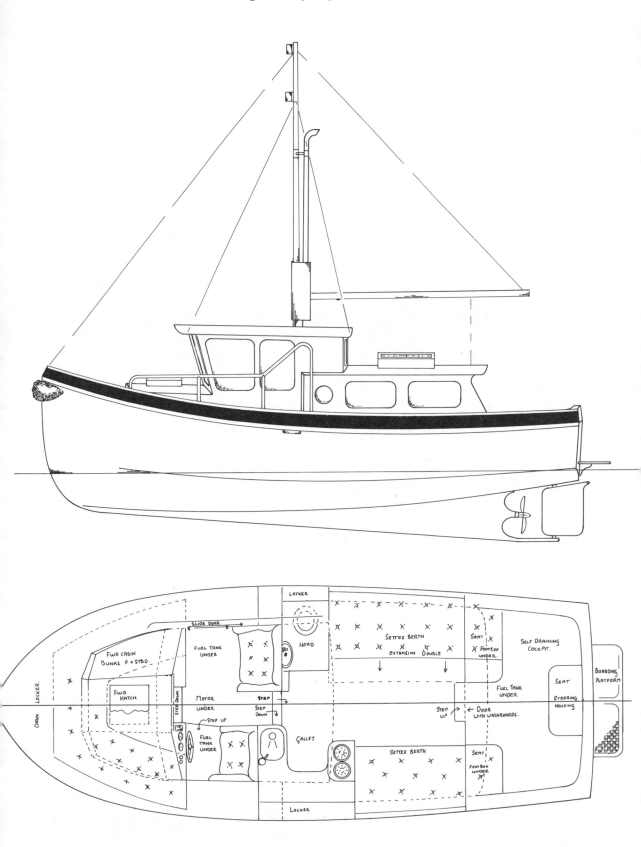

Design Twenty-One: Rainbow Riverboat

A cruising houseboat for private or charter use

Around mid-1992 I was approached by Rainbow Yacht Charters to draw up a river houseboat for a proposed charter fleet. As someone with a fairly open mind about form and structure, but little experience in houseboats or bigger boats on rivers, I started off with a series of study proposals. Each one went through an evaluation process with the Rainbow staff, the resulting suggestions being incorporated in the next drawing.

Dealing with the likes and dislikes of several people at once was a bit disconcerting. Although the process was a very rushed one involving many midnights (oh, the joys of being self-employed), we had the final drawings ready for presentation at the IMTEC boat show that year, and I think the end result was a better boat.

I had been able to draw on the extensive charter experience of Evelyn and Roger Miles to design an interior that would suit their clientele, and with the assistance of naval architect Albert Sedlmeyer and engineer/consultant Richard Downs-Honey, produced an efficient hull form and structure that would be strong enough to cope with coastal (salt water) cruising in addition to the original brief, and which could be production-built to get a fleet up and running quickly.

Rainbow wanted to build the boats of fibreglass, mostly for customer acceptance and low maintenance. This would have been okay, I guess, for volume production, but the projected costs were startling. I would have built the pontoons from steel, and the superstructure from steel-framed ply or foam-backed aluminium sheet, in much the same way as one-off truck bodies. My own costings suggested that a boat built in this manner would have come in at about half the cost!

A fast shoal-draft vessel such as this, particularly one with such a huge interior, lends itself to a lifestyle that few other craft could aspire to. She has convertible double berths in cabins fore and aft, four bunks in a separate bunkroom, wardrobes and cupboards aplenty, and a kitchen at least as good as the one at home. With such a stable craft, domestic appliances can be used throughout the boat. These include pressure hot and cold water, cabin heaters, shower, fridge-freezer, and a wastemaster, with all waste discharging into holding tanks.

Extensive deck areas complement the spacious interior. There is a deck at both bow and stern, plus a complete upper deck with an outside control station 'upstairs'. There is room for the ten people that the Rainbow Riverboat sleeps to each find a quiet corner for those gentle lazy afternoons at anchor, or while on passage up the coast.

This version, powered by a pair of Yanmar D37 diesel outboard motors in soundproofed boxes aft, should be whisper quiet. It should slip along at up to ten knots, use fuel at a miserly rate, and be very economical to build without the complication of inboard motors.

Structurally, this 'boat' is very simple: a pair of square-section pontoons with well-shaped ends, water and waste tanks built in, a galvanised steel connecting structure, and a stylised 'box' with internal stiffening bulkheads which forms the accommodation — simple, effective, and more attractive than many of the overseas examples which we studied at the beginning of the project.

This was one of the good ideas that didn't happen. Alas, a law intended to conserve the country's natural assets put a long-term hold on the project.

This leaves us with a mental picture of how it would be ... to pull up to the riverbank in the middle of Hamilton city and take the kids to

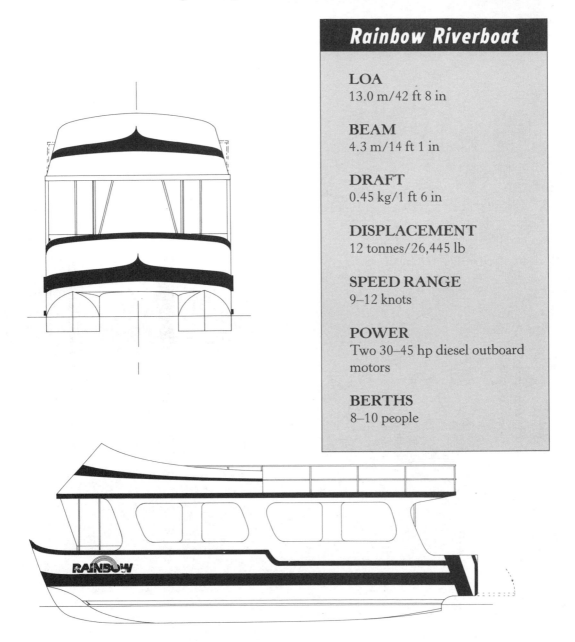

Rainbow Riverboat

LOA
13.0 m/42 ft 8 in

BEAM
4.3 m/14 ft 1 in

DRAFT
0.45 kg/1 ft 6 in

DISPLACEMENT
12 tonnes/26,445 lb

SPEED RANGE
9–12 knots

POWER
Two 30–45 hp diesel outboard motors

BERTHS
8–10 people

McDonalds while the other adults go shopping; to cruise quietly upstream afterwards, through farmland and willow trees, as the sun begins to cast long shadows; to drop the mooring line over a pile and repair to the afterdeck for dinner as the younger crew are tucked into bed; to fish off the stern platform while the others lounge in the forward stateroom; to watch the light among the trees and on the water as the moon comes up, almost heart-stopping in its intensity; to boom down river the next day at ten knots, with the current adding more speed still, heading for the spectacular beach at the mouth of the river; to return more quietly late in the afternoon after a day spent playing cricket, swimming and exploring.

It's no wonder that houseboats are so popular in Australia and the USA; rather, it is a wonder that we do not see more of them here. A coastal-capable displacement catamaran houseboat, such as the *Rainbow Riverboat*, would have to be a great alternative to the bach at the beach!

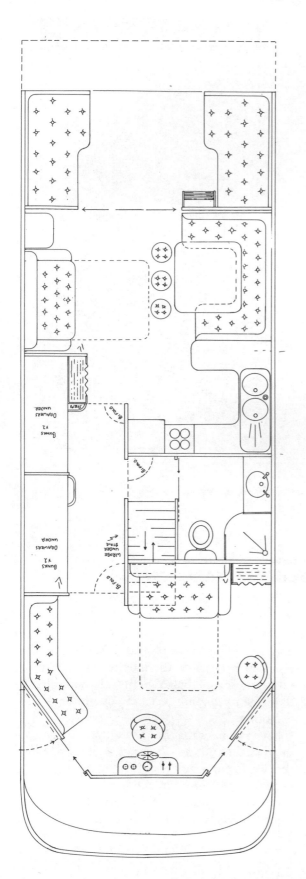

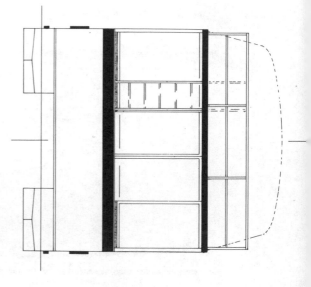

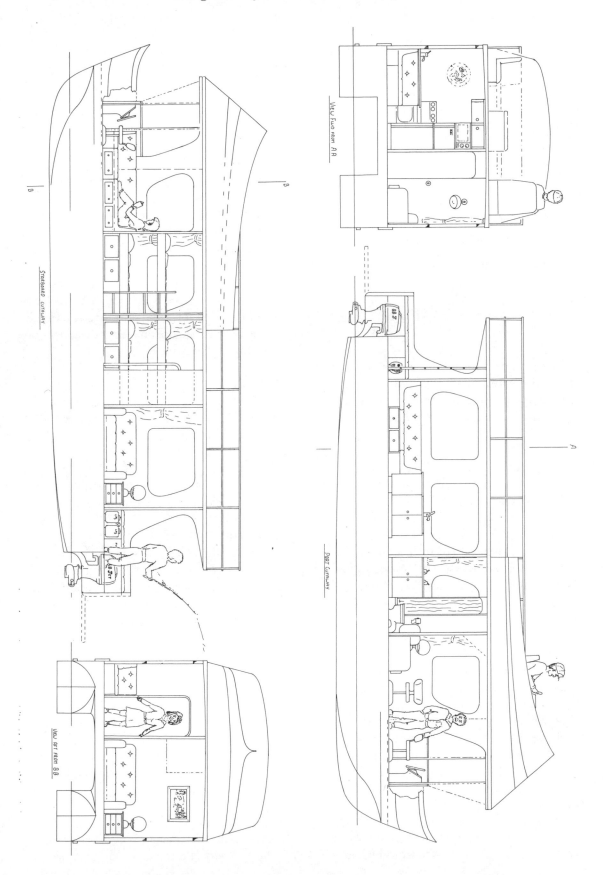

Glossary

Bach
A Welshman's friend; a Kiwi's holiday home; or a spelling mistake when I'm trying to say - - - -.

Bang
Noise used for starting yacht races.

Batch
A mix, as in concrete, glue or scones.

Batten
A 'ruler' for drawing 'fair curves'.

Baulk
What your partner does when you announce your intention to build yet another boat. On a more positive note, a large piece of rough-sawn timber.

Bevel
An angle planed on a piece of the boat's frame so that another part of the structure will lie correctly on it.

Blueing
A discolouration caused by heat, in our case when an edge tool such as a chisel has been ground too aggressively on a power grinder. It will soften the steel and may ruin the tool.

Bottlescrew
Or rigging screw. A device with a left-handed threaded rod in one end and a right-handed threaded rod in the other, so that when the centre section is turned, the ends are drawn together or forced apart. Used for tensioning standing rigging. (Nothing to do with uncorking wine.)

Bow
The sharp end of the boat, except for boats like *Roof Rack* (which doesn't have a sharp end) and the Six-metre Whaler (which is sharp at both ends). Whatever the design, the bow is the front end.

Breast hook
As a young yachtie, I would have made some comment about this being an instrument to undo a certain type of undergarment, but today it is a brace fitted between two gunwales immediately aft of the stem.

Bulkhead
A wall across a boat. Although a structural item serving the same purposes as a frame, it is used to enclose an area for waterproof storage or buoyancy.

Bulwark
A sort of low wall around the gunwale which helps to stop feet sliding over the edge on a rough day.

Butt block
A reinforcing piece used to 'double up' behind a join in the planking.

Butt strap
Also tingle. Different terms for a butt block. (The language of boatbuilding is designed to make those who speak it sound like practitioners of an arcane and complex art. Master the terminology and no one will be able to contradict you as they won't know what on earth you're on about!)

Carvel planking
The traditional way of building a smooth-skinned boat. Uses planks carved to shape, steamed or sawn frames to hold them together, and usually some form of thread (caulking) driven between the planks to keep the water out.

Centreboard
Also drop keel, dagger board, or swing keel. A sort of retractable wing that slows

the rate at which the boat slides sideways when pushed by the wind. A bigger boat has a fixed keel which fulfils the same purpose, as well as carrying weight (ballast) which helps reduce the boat's tendency to lean over when the wind blows on the sails.

Centrecase
The housing into which the centreboard fits when retracted.

Centreline
A line down the middle of the boat from bow to stern. Plan view only (from above).

Checks
Little splits in the surface of the wood or plywood.

Chine
A hard angle in the boat's side when running the eye from the gunwale to the keel. Usually evident in a boat built of sheet material, such as plywood or steel.

Coaming
A little raised wall around an opening such as the seating area (cockpit) or hatch.

Cold moulded
A method of glueing multiple layers of thin woods together to produce a round-sided boat.

Compound curves
A round-sided boat has panels which curve in two directions at once. A chine boat has panels which curve in one 'plane' only.

Conical projection
By using sections of an imaginary cone, it is possible to calculate shapes that can be achieved by wrapping flat sheets around a form. 'Multiconic projection' is a more complex version of the same thing, using sections of cones with differing radii, and tapers to achieve the desired shape.

Crossband
The layer of plywood with the grain going

across the sheet is a 'crossband', as opposed to a 'longband' which runs lengthwise along the long axis of the sheet. Almost always at 90 degrees to each other.

Cutter rig
Similar to a sloop rig, but with two or more jibs and usually a bowsprit. Could have a gaff- or lug-rigged mainsail and mizzen.

Deadrise
The angle at which the bottom rises from the keel (measuring across the boat) forming a 'vee' shape.

Delamination
A coming apart, usually through either overloading or adhesive failure, of a component made of several layers of material.

Dry fit
Your partner's next reaction after baulking! Or, to assemble an item without glue or permanent fastenings (dry), used to check the 'fit'.

Egg-beater
A very small motor — as in outboard. Or a hand drill.

Extender
An additive which, when mixed with epoxy resin, 'extends' the resin's use to become a high-density filler, glue, or many other useful products.

Fillet
A method of reinforcing a corner join with a radius of high-density epoxy filler.

Gaff rig
A mainsail with an extra spar along the top edge, making it four-sided. In the 'gaffer's' case, the luff (or front edge) of the sail is secured to the mast, and the forward end of the 'extra spar' (the gaff) has jaws which locate it on the mast. (The design for *Penguin* is an example of a gaff-rigged mainsail.) See also 'lug sail'.

Galv
Abbreviation of 'hot dip galvanised', a system of rustproofing ferrous metals by dipping them into molten zinc.

Gimbal
A mounting or housing pivoted in both horizontal axes so a compass or a stove (the common items so mounted in our craft) will remain level enough to use during bad weather.

Glue line
The film of glue between two pieces of wood glued together.

Grain
Wood is made up of long spindle-shaped cells, all oriented in a common direction. This direction is known as the 'grain'. Wood generally splits, planes or chisels much more easily along the grain. The term is also used to describe the pattern in the wood which shows up particularly when varnished.

Grounds
Pieces of wood used to reinforce a joint.

Gunwale
Originally a heavy strip of timber (the 'wale') used to protect the gunports on a sailing warship. Nowadays, it refers to the structure around the boat's top edge. Also gun'l or gunnel.

Jig
A system of temporary framing that does not become part of the boat's structure. Not an Irish dance, or a method of fishing with a hand line.

Keelson
Structural member running from stem to stern inside the boat's skin, on the centreline on the bottom.

Ketch rig
Another rig; a ketch has two masts, the front one the bigger with the back one ahead or forward of the rudder. (See also yawl rig.)

Laminate
To make a curved component by glueing a number of thin layers together.

Lamination
The process of making the above.

Lapstrake
Traditional method of boatbuilding which involves overlapping 'strakes' of planking, giving a 'weatherboard' appearance; used for Granddad's clinker-built dinghy.

Longband
See crossband.

Lug sail
Looks quite similar, from a distance, to a gaff sail but has no jaws on the extra spar (or 'yard' in this case) and the sail is not secured to the mast. Usually the luff lies along the mast and may project a little forward. (The designs for *Rogue* and *Houdini* provide examples of lug sails.)

Mizzen
The small sail at the back on a two-masted boat such as a yawl or ketch. On a single-masted craft, it is the sail that is 'mizzen'.

Monel
A high-grade marine alloy, used mostly for fastenings.

Navel pipe
A pipe leading anchor rope or chain from the foredeck into its stowage locker. (Not, as my editor thought, a bizarre new method of smoking!)

Ply
(Plural: plies.) Layers in plywood, otherwise known as veneers.

Quick-and-dirty
A euphemism for 'use every possible shortcut as long as it still keeps the water out!'

Radius
To round off.

Reef
To make a sail smaller, usually to cope with an increase in wind speed. Can also refer to rocks just under the water's surface, or a method of ripping the caulking out of seams of a carvel-built boat. However, you are going to avoid the rocky type like the plague, and you won't be stuffing the leaky parts of your boat with tarred cotton. (Bog it up with filler and fibreglass instead!)

Rig
The mast, sails and their arrangements. The subject of rigs, sail types and their variations is a chapter, if not a whole book, in itself. I'll spare you this time! See cutter, gaff, lug, sloop and yawl rig definitions for examples.

Rocker
The amount of fore and aft curve in the boat's bottom.

Roove
A conical copper washer which forms the head of the rivet used in building the old clinker-style planked dinghy.

Running rigging
See standing rigging.

Scarf join
A method of increasing the glueing surface in a joint by cutting matching slopes in the parts to be joined.

Scarfing
Doing the above.

Set of offsets
Otherwise known as a 'table of offsets', this is a set of measurements which describe the shape of either the boat's hull, or individual components.

Sheer
Or sheerline. The curve (or lack of it) which describes the top edge of the hull when viewed from the side (profile view).

Shoe
A reinforced area at the tip of a keel, or along the centre of a speedboat's bottom.

Skeg
A shallow keel at the after end of the boat. Usually used to improve a boat's directional stability.

Sloop rig
Yet another rig! A boat that is sloop rigged has one mast, with one sail ahead of it (the jib) and one behind (the main, or mainsail). Could have a gaff- or lug-rigged mainsail.

Splashguard
A small extension of the foredeck out over the gunwale. Used to reduce the spray that comes over the bows in rough weather.

Splice
To join.

Sprit boom
A 'sprit' is a spar or pole used to hold a sail out. In the case of our little boats, look at the designs for *Rogue* and *Houdini*, which demonstrate the meaning better than I could explain in one line or less. Note how the boom's forward end is higher than the bottom of the sail; this means the usual vang, or kicking strap, is not needed.

Standing rigging
The stays which hold the mast up, as opposed to 'running rigging', the ropes used to hoist and control the sail(s).

Stem
Structural member forming the shape of the boat's bow, and joining the two sides together.

Sternsheets
Traditional term for the seats at the back of an open or partly decked boat.

Strake
A continuous line of planking from the boat's stem to its stern. The sides of boats such as *Navigator* and *Tender Behind* are formed of several 'strakes' of planking. May be a derivative of 'streaks' which is an indication of the appearance. See lapstrake.

Stress riser
An engineering term used to describe the situation where a sudden change of size can weaken a component.

Stringer
A lengthwise piece of framing, usually running the full length of the boat. The frames and bulkheads form the crosswise items which hold the stringers in position.

Tabernacle
Certain religious types take one of these aboard so as not to be caught short if not back by their holy day. The rest of us use one to step the base of the mast in, allowing the mast to be more easily raised and lowered.

Table of offsets
See set of offsets.

Tanalised
(Tanalith™.) Trade name for a process of pressure-treating timber with a preservative based on copper, chrome and arsenic. The active ingredients are chemically bonded to the wood's cellular structure and cannot leach out.

Taper
A reduction in thickness towards one end: a cone 'tapers' from one end to another.

Tarp
Abbreviation for tarpaulin, or canvas cover. Today it seems to refer to any large rectangle of waterproof fabric.

Thwartships
At 90 degrees from the centreline in plan view, or 'across the boat'.

Tingle
See butt strap.

Tortured ply
A building method of forcing plywood into compound shapes; the term 'tortured' may also apply to the boatbuilder who tries it.

Transom
The flat piece across the back of the boat.

Trim
Not a way of making sails smaller when the breeze pipes up, but a reference to either the fore and aft balance of the boat in terms of weight, or, in the case of keen racers, 'trim (as in adjust) that sail' (and look lively about it!)

Well deck
Why not? Sailing is a very healthy sport! Otherwise, an area of deck lowered to either stow gear, such as anchors, or provide better footing for working the vessel.

Yawl rig
Sorry, last one! A yawl has two masts, the front one the bigger (same as a ketch), but the back one is mounted aft of the rudder post; or in the case of a small boat, with most of the sail behind the rudder.

Disclaimer
In amongst the definitions above will be some that are not historically correct. However, our language is a living one and subject to evolution. There are features on modern boats which defy description by other means, so common usage has us slightly misusing the age-old terminology. If you don't like it, write to my publishers; they are under strict instructions as to where they should put such letters. J.W.

Buying Your Plans

For those who might like to have a try at boatbuilding, a set of plans is the first step. Plans for most of the boats described in this book are available from the author. (The following prices are in New Zealand dollars, and include GST.)

Roof Rack The littlest yacht tender
$65.00

Tender Behind The biggest little boat
$75.00

Fish Hook Utility dinghy
$75.00

Setnet/Golden Bay Dinghy For fishing or sailing
$85.00

Seagull A simple but sporting rowboat
$85.00

Joansa The two oarspower cruiser
$85.00

Light Dory Seriously seaworthy rowboat
$85.00

Daniel's Boat A young man's joy
$95.00

Janette Sailing or fishing for the family
$110.00

Rogue Traditional cruising dinghy with sneaky speed
$125.00

Houdini Daysailer with cruising ability
$125.00

Navigator Club racer or knockabout cruiser
$150.00

Sweet Pea Lightweight trailer yacht
$195.00

Six-Metre Whaler Sailtrainer or big daysailer with a navy flavour
$185.00

Penguin Capable coastal cruising trailer yacht
$275.00

Rifleman Lightweight outboard runabout
$125.00

Plans for the other craft shown in the text may be available, price on application. An up-to-date catalogue, which includes plans completed since this book was published, is available for $15.00 (post paid) from:

John Welsford Small Craft Design,
PO Box 314,
Ngongotaha, Rotorua, New Zealand.
Ph/Fax: +64 (0)7 3575354 (after hours), email: jwboatdesigns@usa.net

All plans are delivered post free within New Zealand.

If ordering in Australia, the prices remain the same, but in Australian dollars (the difference pays for the airmail) or order from the Australian agent:

West Tamar Wooden Boats,
601 Devoit Road,
Devoit,
Tasmania 7275,
Australia.
Ph: (61) (03) 6394 7165 Fax: (61) (03) 6394 7168
email: wtwb@tassie.net.au Website: http://www.taswebcom.au/wtwb/index.htm

For other countries, please convert the price at the going rate (phone the bank — they'll have it) and add 15 percent for airmail postage. A cheque in your own currency is fine.

Thank you for your interest. I hope you've enjoyed your reading.

Fair winds and a sheltered anchorage.

John Welsford